Introduction to Matrix Analysis and Applications

Introduction to Matrix Analysis and Applications

Editor

Mohamed Ali Karim

Introduction to Matrix Analysis and Applications

Edited by **Mohamed Ali Karim**

Printed in 2017

ISBN: 978-1-68117-187-6

Library of Congress Control Number: 2015949132

© 2016 by
SCITUS Academics LLC,
616, Corporate Way, Suite 2, 4766,
Valley Cottage, NY 10989

www.scitusacademics.com

Preface

Matrix analysis is the study of matrices and their algebraic properties. Some particular topics out of many include; operations defined on matrices (such as matrix addition, matrix multiplication and operations derived from these), functions of matrices (such as matrix exponentiation and matrix logarithm, and even sines and cosines etc. of matrices), and the eigenvalues of matrices (eigendecomposition of a matrix, eigenvalue perturbation theory).

Matrices can be studied in different ways. They are a linear algebraic structure and have a topological/analytical aspect and they also carry an order structure that is induced by positive semidefinite matrices. The interplay of these closely related structures is an essential feature of matrix analysis. The book treats some aspects of analysis related to matrices including such topics as matrix monotone functions, matrix means, majorization, entropies, quantum Markov triplets. There are several popular matrix applications for quantum theory.

This book emphasizes on these aspects of matrix analysis from aneffient analysis point of view. Applications include such areas as signal processing, systems and control theory, statistics, Markov chains, and mathematical biology. Introduction to Matrix Analysis and Applications is appropriate for an advanced graduate course on matrix analysis. It can also be used as a reference for researchers in quantum information, statistics, engineering and economics.

Contents

Matrix Measure with Application in Quantized Synchronization Analysis of Complex Networks with Delayed Time via the General Intermittent Control

Qunli Zhang
Department of Mathematics,
Heze University, Heze, China

1

ABSTRACT

This paper concerned with the quantized synchronization analysis problem. The scope of state vectors of dynamic systems, based on the matrix measure, is estimated. By using the general intermittent control, some simple yet generic criteria are derived ensuring the exponential stability of dynamic systems. Then, both the general intermittent networked controller and the quantized parameters can be designed, which guarantee that the nodes of the complex network are synchronized. Finally, simulation examples are given to illustrate the effectiveness and feasibility of the proposed method.

INTRODUCTION

Since its origins in the work of Fujisaka and Yamada [1- 3], Afraimovich, Verichev, and Rabinovich [4], and Pecora and Carrol [5], the study of synchronization of chaotic systems [6-19] is of great practical significance and has received great interest in recent years. In the above literatures, the approach applied to stability analysis is basically the Lyapunov's method. As we all know, the construction of a proper Lyapunov function usually becomes very skillful, and the Lyapunov's method does not specifically describe the convergence rate near the equilibrium point of the system. Hence, there is little compatibility among all of the stability criteria obtained so far.

The concept named the matrix measur [20-25] has been applied to the investigation of the existence, uniqueness or stability analysis of the equilibrium. Intermittent control [26-29] has been used for a variety of purposes in engineering fields such as manufacturing, transportation, air-quality control and communication. A wide variety of synchronization or stabilization using the periodically intermittent control method has been studied (see [27- 32]). Compared with continuous control methods [7-14], intermittent control is more efficient when the system output is measured intermittently rather than continuously. All of intermittent control and impulsive control are belong to switch control. But the intermittent control is different from the impulsive control, because impulsive control is activated only at some isolated moments, namely it is of zero duation, while intermittent control has a nonzero control width.

But it should be mentioned that the influence caused by quantization has not been considered in their results. It is well known that in modern networked systems, quantization is an indispensable step that aims at saving limited bandwidth and energy consumption [33]. Quantization cannot be avoided in the digital control setting, and it is indeed a natural way to be inserted into the control design complexity constraints of the controller and communication constraints of the channels which connect the controller and the plant [34]. The important application of quantization in real world can be found in human machine interaction, for instance, see [35-37]. Therefore, it is essential and important to investigate the exponential quantized synchronization problem of networks with mixed delays by periodically intermittent control.

Our interest focuses on the class of commonly intermittent controller with time duration, where the control is activated in certain nonzero time intervals, and is off in other time intervals. A special case of such a control law is of the form

$$U(t) = \begin{cases} -k\big(y(t) - x(t)\big), & (nT \le t < nT + \delta), \\ 0, & (nT + \delta \le t < (n+1)T), \end{cases}$$

Where k denotes the control strength, $\delta > 0$ denotes the switching width, and T denotes the control period. The general intermittent controller

$$U(t) = \begin{cases} -k\big(y(t) - x(t)\big), & \big(h(n)T \le t < h(n)T + \delta\big), \\ 0, & \big(h(n)T + \delta \le t < h(n+1)T\big), \end{cases}$$

Where h(n) is a strictly monotone increasing function on n , has been studied (see [38]).

Moreover, a logarithmic quantizer q(α) has quantization levels given by

$\Pi = \{\pm\alpha_c \,|\, \alpha_c = \rho^c \mu_0, c = 0, \pm1, \pm2, ...\} \cup \{0\}$, Where the quantization densitie is $\rho \in (0,1)$, and the scaling parameter is $\mu_0 > 0$. Then, the quantizer q(α) is defined as follow

$$q(\alpha) = \begin{cases} \rho^c \mu_0, & \text{if } \dfrac{\rho^c \mu_0}{1+\beta} < \alpha \le \dfrac{\rho^c \mu_0}{1-\beta} \\ 0, & \text{if } \alpha = 0 \\ -q(-\alpha), & \text{if } \alpha < 0 \end{cases}$$

$$(1)$$

Where $\beta = \dfrac{1-\rho}{1+\rho}$ Based on (1), it is obvious that $q(\alpha) - \alpha = \Delta\alpha$ and the quantization synchronization error $\Delta \in [-\beta, \beta]$ (see [39-43]).

In this paper, based on matrix measure and Gronwall inequality, the general intermittent controller

$$U(t) = \begin{cases} -kq\big(x_i(t) - s(t)\big), & \big(h(n)T \le t < h(n)T + \delta\big), \\ 0, & \big(h(n)T + \delta \le t < h(n+1)T\big), \end{cases}$$

Where h(n) is a strictly monotone increasing function on n,

$$U(t)$$
$$= \begin{cases} -kq\big(x_i(t) - s(t)\big), & \big(h(n+1)T \le t < h(n+1)T + \delta\big), \\ 0, & \big(h(n+1)T + \delta \le t < h(n)T\big), \end{cases}$$

Where h(n) a strictly monotone decreasing function on n is, is designed. Then the sufficient yet generic criteria for synchronization of complex networks with and without delayed item are obtained.

This paper is organized as follows. In Section 2, some necessary background materials are presented. In Section 3, the state vectors scope estimated via matrix measure are formulated. Section 4 deals with the quantized synchronization. The theoretical results are applied to complex networks, and numerical simulations of delayed neural network systems are shown in this section. Finally, some concluding remarks are given in Section 5.

PRELIMINARIES

Let X be a Banach space endowed with the l²-norm

$$\| \|_{ie}, \|x\| = \sqrt{x^T x} = \sqrt{\langle x, x \rangle},$$ where $\langle , \rangle$ is inner product, and Ω be a open subset of X. We consider the following system:

$$\frac{dx}{dt} = -Cx(t) + F\big(x(t)\big) + G\big(x(t-\tau)\big), \tag{2}$$

Where F,G are nonlinear operators defined on Ω, and $x(t), x(t-\tau) \in \Omega$, and τ is a time-delayed positive constant, and $F(0) = G(0) = 0$.

Definition 1: [12, 26, 28, 44] System (2) is called to be exponentially stable on a neighborhood Ω of the equilibrium point, if there exist constants $\mu > 0, \alpha > 0$, such that $\|x(t)\| \le \alpha \exp(-\mu t)\|x_0\|$ $(t > 0)$, Where $x(t)$ is any solution of (2) initiated from $x(t_0) = x_0$.

Definition 2: [20-25] suppose that $M \in R^{n \times n}$ is a matrix. Let $\mu(M)$ be the matrix measure of M defined as

$$\mu(M) = \lim_{\delta \to 0^+} \frac{\|I + \delta M\| - I}{\delta} \text{ Where I is the identity matrix.}$$

Lemma: [20-25] the matrix measure $\mu(M)$ is well defined for the l^2-norm $\|x\| = \sqrt{x^T x} = \sqrt{\langle x, x \rangle}$, the induced matrix measure is given by

$$\mu(M) = \max_i \left(\frac{\lambda_i(M + M^T)}{2} \right), \text{ where } \lambda_i(M + M^T) \text{ denotes all eigenvalues}$$
of the matrix $M + M^T$.

ESTIMATING THE SCOPE OF THE STATE VECTORS

We consider the following system:

$$\frac{dx}{dt} = -Cx(t) + f\big(x(t), y(t)\big) + g\big(x(t-\tau)\big), \tag{3}$$

$$\frac{dy}{dt} = -Cy(t) + m\big(x(t), y(t)\big) + g\big(y(t-\tau)\big), \tag{4}$$

Where f, m, g are nonlinear operators defined on Ω, and $x(t), y(t)$ $x(t-\tau), y(t-\tau) \in \Omega$, and τ is a time delayed positive constant, and $f(0,0) = m(0,0) = g(0) = 0$.

Theorem 1: For any $x, y, \in \Omega$ in the system (3), (4), if the operator f, m, g satisfies

$$\|g(x) - g(y)\| \le l\|x - y\|,$$

$$(5)$$

f,m is bound, where l is a positive constant. The solutions $x(t), y(t)$, initiated from $x(t_0) = x_0 \in \Omega, y(t_0) = y_0 \in \Omega$ of the system (3) and (4) satisfy

$$\|x - y\| \le \|x_0 - y_0\| \exp\{\rho(t - t_0)\}, \quad \forall t \ge 0,$$

where $\rho = \eta + le^{-\eta 1\tau}, \eta = \mu(-C) + \lambda$,

$$\lambda = \max_{\forall t \ge 0} \frac{\left(x(t) - y(t)\right)^T \left(f\left(x(t), y(t)\right) - m\left(x(t), y(t)\right)\right)}{\|x(t) - y(t)\|^2}.$$

Proof: Under the initial conditions $x(t_0) = x_0 \in \Omega, y(t_0) = y_0 \in \Omega$ we have

$$\frac{d\|x(t) - y(t)\|}{dt} - \mu(-C)\|x(t) - y(t)\|$$

$$= \lim_{\delta \to 0^+} \frac{1}{\delta}\left(\|x(t+\delta) - y(t+\delta)\| - \|I + \delta(-C)\|\|x(t) - y(t)\|\right)$$

$$\le \lim_{\delta \to 0^+} \frac{1}{\delta}\left(\|x(t+\delta) - y(t+\delta)\|\right.$$

$$\left. - \|(I + \delta(-C))(x(t) - y(t))\|\right)$$

for any $t \ge 0$.

Let $u(t,\delta) = x(t+\delta) - y(t+\delta)$,

$$v(t,\delta) = \left(I + \delta(-C)\right)\left(x(t) - y(t)\right)$$
$$= \left(I + \delta(-C)\right)u(t,0),$$

Then

$$\frac{d\|x(t)-y(t)\|}{dt} - \mu(-C)\|x(t)-y(t)\| \le \lim_{\delta \to 0^+} \frac{1}{\delta}\left(\sqrt{u^T(t,\delta)u(t,\delta)} - \sqrt{v^T(t,\delta)v(t,\delta)}\right)$$

$$= \lim_{\delta \to 0^+} \frac{1}{\delta} \frac{u^T(t,\delta)u(t,\delta) - v^T(t,\delta)v(t,\delta)}{\sqrt{u^T(t,\delta)u(t,\delta)} + \sqrt{v^T(t,\delta)v(t,\delta)}} = \lim_{\delta \to 0^+} \frac{1}{\|u(t,\delta)\| + \|v(t,\delta)\|} \lim_{\delta \to 0^+} \frac{1}{\delta}\left(u^T(t,\delta)u(t,\delta) - v^T(t,\delta)v(t,\delta)\right)$$

$$= \frac{1}{2\|x(t)-y(t)\|}\left(\lim_{\delta \to 0^+} \frac{u^T(t,\delta)u(t,\delta) - u^T(t,0)u(t,0)}{\delta} + u^T(t,0)\left(C+C^T\right)u(t,0) + \lim_{\delta \to 0^+} \delta u^T(t,0)\left(C^TC\right)u(t,0)\right)$$

$$= \frac{1}{2\|x(t)-y(t)\|}\left(\left[\frac{d(u^T(t,0)u(t,0))}{dt} + u^T(t,0)\left(C+C^T\right)u(t,0)\right] = \frac{1}{2\|x(t)-y(t)\|}\left(\frac{d\left((x(t)-y(t))^T(x(t)-y(t))\right)}{dt}\right.\right.$$

$$\left. + (x(t)-y(t))^T\left(C+C^T\right)(x(t)-y(t))\right) = \frac{1}{\|x(t)-y(t)\|}(x(t)-y(t))^T\left(f(x(t),y(t)) - m(x(t),y(t))\right)$$

$$+ (x(t)-y(t))^T\left(g(x(t-\tau)) - g(y(t-\tau))\right)$$

Using Cauchy-Bunyakovsky Inequality and condition (5), we obtain

$$\frac{d\|x(t)-y(t)\|}{dt} - \mu(-C)\|x(t)-y(t)\| \le \frac{(x(t)-y(t))^T\left(f(x(t),y(t)) - m(x(t),y(t))\right)}{\|x(t)-y(t)\|} + \|g(x(t-\tau)) - g(y(t-\tau))\|$$

$$\le \frac{(x(t)-y(t))^T\left(f(x(t),y(t)) - m(x(t),y(t))\right)}{\|x(t)-y(t)\|^2} \cdot \|x(t)-y(t)\| + l\|x(t-\tau) - y(t-\tau)\|$$

$$\le \lambda\|x(t)-y(t)\| + l\|x(t-\tau) - y(t-\tau)\|.$$

So

$$\frac{d\|x(t)-y(t)\|}{dt} \le \left(\mu(-C) + \lambda\right)\|x(t)-y(t)\|$$
$$+ l\|x(t-\tau) - y(t-\tau)\|.$$

$$\left\| x(t) - y(t) \right\| \le \left\| x_0 - y_0 \right\| e^{(\mu(-C) + \lambda)(t - t_0)}$$

$$+ \int_{t_0}^{t} e^{(\mu(-C) + \lambda)(t - t_0)} l \left\| x(s - \tau) - y(s - \tau) \right\| ds,$$

Namely

$$e^{-\eta(t - t_0)} \left\| x(t) - y(t) \right\|$$

$$\le \left\| x_0 - y_0 \right\| + \int_{t_0}^{t} l e^{\eta(s - t_0)} \left\| x(s - \tau) - y(s - \tau) \right\| ds$$

$$= \left\| x_0 - y_0 \right\| + l e^{-\eta \tau} \int_{t_0 - \tau}^{t - \tau} e^{-\eta(s - t_0)} \left\| x(s) - y(s) \right\| ds.$$

Using the Gronwall inequality [45, 46], we have

$$e^{-\eta(t - t_0)} \left\| x(t) - y(t) \right\| \le \left\| x_0 - y_0 \right\| \exp \left\{ l e^{-\eta \tau} (t - t_0) \right\},$$

That is

$$\left\| x(t) - y(t) \right\| \le \left\| x_0 - y_0 \right\| \exp \left\{ \left(\alpha + l e^{-\eta \tau} \right)(t - t_0) \right\}$$

$$\le \left\| x_0 - y_0 \right\| \exp \left\{ \rho(t - t_0) \right\}.$$

SYNCHRONIZATION VIA THE GENERAL INTERMITTENT CONTROL AND EXAMPLES

Consider a delayed complex dynamical network consisting of N linearly coupled no identical nodes described by

$$\frac{dx_i(t)}{dt} = -Cx_i(t) + p(x_i(t)) + g(x_i(t-\tau))$$

$$+ \sum_{j=1}^{N} a_{ij} x_j(t) + u_i(t), \quad i = 1, 2, \cdots, N, \tag{6}$$

where $x_i = (x_{i1}, x_{i2}..., x_{in})^T \in R^n$ is the state vector of the ith node, $p, g : R^n \to R^n$ are nonlinear vector functions, $u_i(t)$ is the control input of the ith node, and $A = (\alpha_{ij})_{N \times N}$ is the coupling figuration matrix representing the coupling strength and the topological structure of the complex networks, in which $\alpha_{ij} > 0$ if there is connection from node i to node $j(i \neq j)$, and is zero, otherwise, and the constraint

$$a_{ii} = -\sum_{j=1, j \neq i}^{N} a_{ij} = -\sum_{i=1, i \neq j}^{N} a_{ij}, \quad (i, j = 1, 2, \cdots, N),$$

is set.

A complex network is said to achieve asymptotical synchronization if

$$x_1(t) = x_2(t) = \cdots = x_N(t) = s(t) \text{ as } t \to \infty \tag{7}$$

Where $s(t) \in R^n$ is a solution of a real target node, satisfying?

$$\frac{ds(t)}{dt} = -Cs(t) + p(s(t)) + g(s(t-\tau)).$$

For our synchronization scheme, let us define error vector and control input $u_i(t)$ as follows, respectively:

$e_i(t) = x_i(t) - s(t), i = 1, 2..., N$. When $h(n)$ is a strictly monotone increasing function on n with $h(0) = 0, \lim_{n \to \infty} h(n) = +\infty,$

$$u_i(t) = \begin{cases} -kq(e_i(t)), & (h(n)T \le t < h(n)T+\delta), \\ 0, & (h(n)T+\delta \le t < h(n+1)T), \end{cases}$$
$$(k > 0, i = 1, 2, \cdots, N) \tag{8}$$

When $h(n)$ is a strictly monotone decreasing function on n with? $h(0) = +\infty$, $\lim_{n \to \infty} h(n) = 0$,

$$u_i(t) = \begin{cases} -kq(e_i(t)), & (h(n+1)T \le t < h(n+1)T+\delta), \\ 0, & (h(n+1)T+\delta \le t < h(n)T). \end{cases}$$
$$(k > 0, i = 1, 2, \cdots, N) \tag{9}$$

In this work, the goal is to design suitable function $h(n)$ and parameters δ, T, and k satisfying the condition (7). The error system follows from the expression (6), (8) and (9)

$$\begin{cases} \dfrac{de_1(t)}{dt} = -C(x_1(t)-s(t))+p(x_1(t))-p(s(t)) \\ \qquad + g(x_1(t-\tau))-g(s(t-\tau))+\sum_{j=1}^{N}a_{1j}e_j(t)+u_1(t), \\ \dfrac{de_2(t)}{dt} = -C(x_2(t)-s(t))+p(x_2(t))-p(s(t)) \\ \qquad + g(x_2(t-\tau))-g(s(t-\tau))+\sum_{j=1}^{N}a_{2j}e_j(t)+u_2(t), \\ \qquad \vdots \\ \dfrac{de_N(t)}{dt} = -C(x_N(t)-s(t))+p(x_N(t))-p(s(t)) \\ \qquad + g(x_N(t-\tau))-g(s(t-\tau))+\sum_{j=1}^{N}a_{Nj}e_j(t)+u_N(t). \end{cases}$$
$$\tag{10}$$

When $h(n)$ is a strictly monotone increasing function on n with $h(0) = 0$, $\lim_{n \to \infty} h(n) = +\infty$, we obtain the following result:

$$\frac{dx_i(t)}{dt} = -Cx_i(t) + p(x_i(t)) + g(x_i(t-\tau))$$

$$+ \sum_{j=1}^{N} a_{ij} x_j(t) + u_i(t), \quad i = 1, 2, \cdots, N.$$

$$(6)$$

where $x_i = (x_{i1}, x_{i2}..., x_{in})^T \in R^n$ is the state vector of the ith node, $p, g : R^n \to R^n$ are nonlinear vector functions, $u_i(t)$ is the control input of the ith node, and $A = (\alpha_{ij})_{N \times N}$ is the coupling figuration matrix representing the coupling strength and the topological structure of the complex networks, in which $\alpha_{ij} > 0$ if there is connection from node i to node $j (i \neq j)$, and is zero, otherwise, and the constraint

$$a_{ii} = -\sum_{j=1, j \neq i}^{N} a_{ij} = -\sum_{i=1, i \neq j}^{N} a_{ij}, \quad (i, j = 1, 2, \cdots, N),$$

is set.

A complex network is said to achieve asymptotical synchronization if

$$x_1(t) = x_2(t) = \cdots = x_N(t) = s(t) \text{ as } t \to \infty$$

$$(7)$$

Where $s(t) \in R^n$ is a solution of a real target node, satisfying?

$$\frac{ds(t)}{dt} = -Cs(t) + p(s(t)) + g(s(t-\tau)).$$

For our synchronization scheme, let us define error vector and control input $u_i(t)$ as follows, respectively:

$e_i(t) = x_i(t) - s(t), i = 1, 2..., N$. When $h(n)$ is a strictly monotone increasing function on n with $h(0) = 0, \lim_{n \to \infty} h(n) = +\infty$,

$$u_i(t) = \begin{cases} -kq\big(e_i(t)\big), & \big(h(n)T \le t < h(n)T+\delta\big), \\ 0, & \big(h(n)T+\delta \le t < h(n+1)T\big), \end{cases}$$
$$\big(k > 0, i = 1, 2, \cdots, N\big) \tag{8}$$

When $h(n)$ is a strictly monotone decreasing function on n with? $h(0) = +\infty$, $\lim\limits_{n \to \infty} h(n) = 0$,

$$u_i(t) = \begin{cases} -kq(e_i(t)), & \big(h(n+1)T \le t < h(n+1)T+\delta\big), \\ 0, & \big(h(n+1)T+\delta \le t < h(n)T\big). \end{cases}$$
$$\big(k > 0, i = 1, 2, \cdots, N\big) \tag{9}$$

In this work, the goal is to design suitable function $h(n)$ and parameters δ, T, and k satisfying the condition (7). The error system follows from the expression (6), (8) and (9)

$$\begin{cases} \dfrac{de_1(t)}{dt} = -C\big(x_1(t) - s(t)\big) + p\big(x_1(t)\big) - p\big(s(t)\big) \\ \qquad + g\big(x_1(t-\tau)\big) - g\big(s(t-\tau)\big) + \sum\limits_{j=1}^{N} a_{1j} e_j(t) + u_1(t), \\[2mm] \dfrac{de_2(t)}{dt} = -C\big(x_2(t) - s(t)\big) + p\big(x_2(t)\big) - p\big(s(t)\big) \\ \qquad + g\big(x_2(t-\tau)\big) - g\big(s(t-\tau)\big) + \sum\limits_{j=1}^{N} a_{2j} e_j(t) + u_2(t). \\[2mm] \qquad \vdots \\[1mm] \dfrac{de_N(t)}{dt} = -C\big(x_N(t) - s(t)\big) + p\big(x_N(t)\big) - p\big(s(t)\big) \\ \qquad + g\big(x_N(t-\tau)\big) - g\big(s(t-\tau)\big) + \sum\limits_{j=1}^{N} a_{Nj} e_j(t) + u_N(t). \end{cases}$$
$$\tag{10}$$

When $h(n)$ is a strictly monotone increasing function on n with $h(0) = 0$, $\lim\limits_{n \to \infty} h(n) = +\infty$, we obtain the following result:

***Theorem* 2:** Suppose that the operator g in the network (6) satisfies condition (5), and $\mu(-C)$ is defined as Definition 2,

$$\rho_1 = \eta_1 + le^{-\eta_1\tau}, \eta_1 = \mu(-C) + \lambda - k(1+\Delta) \qquad \rho_2 = \eta_2 + le^{-\eta_1\tau}, \eta_2 = \mu(-C) + \lambda$$

Where the constant

$$\lambda = \max_{\forall t \geq 0} \frac{\left(x(t) - s(t)\right)^T \left(f\left(x(t), s(t)\right) - m\left(x(t), s(t)\right)\right)}{\left\| x(t) - s(t) \right\|^2},$$

$$f\left(x(t), s(t)\right)$$

$$= \left(\left(p\left(x_1(t)\right) + \sum_{j=1}^N a_{1j} x_j(t)\right)^T, \left(p\left(x_2(t)\right) + \sum_{j=1}^N a_{2j} x_j(t)\right)^T, \right.$$

$$\left. \cdots, \left(p\left(x_N(t)\right) + \sum_{j=1}^N a_{Nj} x_j(t) + u_N(t)\right)^T\right)^T,$$

$$m\left(x(t), s(t)\right)$$

$$= \left(\left(p\left(s_1(t)\right) + \sum_{j=1}^N a_{1j} s_j(t)\right)^T, \left(p\left(s_2(t)\right) + \sum_{j=1}^N a_{2j} s_j(t)\right)^T, \right.$$

$$\left. \cdots, \left(p\left(s_N(t)\right) + \sum_{j=1}^N a_{Nj} s_j(t)\right)^T\right)^T,$$

$l_i l = \max\{l_1, l_2, \dots l_N\}, e(t) = \left(e_1^T(t), e_2^T(t), \dots, e_N^T(t)\right)^T$, Satisfies

$\left\| g(x_i)(t-\tau) - g(s(t-\tau)) \right\| \leq l_i \left\| e_i(t-\tau) \right\|.$

Then the synchronization of network (6) is achieved if the parameters δ, T, k, λ and ζ satisfy

$\rho_1 > 0, \ \rho_2 > 0,$

$$\inf\left[(\rho_1 + \rho_2)\delta\frac{h^{-1}\left((t-\delta)/T\right)}{t} - \rho_2 \right] \geq \zeta > 0,$$

(11)

Where $h^{-1}(\bullet)$ is the inverse function of the function? $h(\bullet)$

Proof: From Theorem 1, the following conclusion is valid:

$$\|e(t)\| \leq \|e(h(n))T\|\exp\{-\rho_1(t - h(n)T)\}$$

(12)

For any $h(n)T \leq t < h(n)T + \delta$;

$$\|e(t)\| \leq \|e(h(n)T + \delta)\|\exp\{\rho_2(t - h(n)T - \delta)\}$$

(13)

For any $h(n)T + \delta \leq t < h(n+1)T$

In the following, we use mathematical induction to prove, for any non-negative integer n,

$$\|e(t)\|$$
$$\leq \begin{cases} \|e(0)\|\exp\{-\rho_1 t + (\rho_1 + \rho_2)h(n)T - n(\rho_1 + \rho_2)\delta\}, \\ \qquad (h(n)T \leq t < h(n)T + \delta), \\ \|e(0)\|\exp\{\rho_2 t - (n+1)(\rho_1 + \rho_2)\delta\}, \\ \qquad (h(n)T + \delta \leq t < h(n+1)T), \end{cases}$$

(14)

1. For $n \leq 0$, from (12) and (13), we can see that

   a. For $h(0)T + \delta \leq t < h(1)T + \delta$,

$$\|e(t)\| \leq \|e(h(0)T)\| \exp\{-\rho_1(t-h(0)T)\}$$
$$= \|e(0)\| \exp\{-\rho_1 t + (\rho_1 + \rho_2)h(0)T - 0 \cdot (\rho_1 + \rho_2)\delta\},$$

$$\|e(h(0)T+\delta)\|$$
$$\leq \|e(h(0)T)\| \exp\{-\rho_1(h(0)T+\delta-h(0)T)\}$$
$$= \|e(0)\| \exp\{-\rho_1\delta\}.$$

a. For $h(0)T + \delta \leq t < h(1)T$,

$$\|e(t)\| \leq \|e(h(0)T+\delta)\| \exp\{\rho_2(t-h(0)T-\delta)\}$$
$$\leq \|e(0)\| \exp\{-\rho_1\delta\} \exp\{\rho_2(t-h(0)T-\delta)\}$$
$$= \|e(0)\| \exp\{\rho_2 t - (\rho_1 + \rho_2)\delta\}$$
$$= \|e(0)\| \exp\{\rho_2 t - (0+1)(\rho_1 + \rho_2)\delta\}.$$

So (14) is true for $n = 0$.

2. Assume that (14) is true for all $n \leq j$, that is

$$\|e(t)\|$$
$$\leq \|e(0)\| \exp\{-\rho_1 t + (\rho_1 + \rho_2)h(j)T - j(\rho_1 + \rho_2)\delta\},$$
$$t \in [h(j)T, h(j)T+\delta),$$

$$\|e(t)\| \leq \|e(0)\| \exp\{\rho_2 t - (j+1)(\rho_1 + \rho_2)\delta\},$$
$$t \in [h(j)T+\delta, h(j+1)T),$$

$$\|e(h(j+1)T)\|$$
$$\leq \|e(0)\| \exp\{\rho_2 h(j+1)T - (j+1)(\rho_1 + \rho_2)\delta\}.$$

We will prove (14) is also true when $n = j + 1$. From (12) and (13), it is easy to see that

$$\|e(t)\| \le \|e(h(j+1)T)\| \exp\{-\rho_1(t - h(j+1)T)\},$$
$$t \in [h(j+1)T, h(j+1)T + \delta),$$
$$\|e(t)\| \le \|e(h(j+1)T + \delta)\| \exp\{\rho_2(t - h(j+1)T - \delta)\},$$
$$t \in [h(j+1)T + \delta, h(j+2)T).$$

Then, for $t \in [h(j+1)T, h(j+1)T + \delta]$, we have

$$\|e(t)\| \le \|e(0)\| \exp\{\rho_2 h(j+1)T - (j+1)(\rho_1 + \rho_2)\delta\} \exp\{-\rho_1(t - h(j+1)T)\}$$
$$= \|e(0)\| \exp\{-\rho_1 t + (\rho_1 + \rho_2)h(j+1)T - (j+1)(\rho_1 + \rho_2)\delta\},$$

$$\|e(h(j+1)T + \delta)\| \le \|e(0)\| \exp\{-\rho_1(h(j+1)T + \delta)$$
$$+ (\rho_1 + \rho_2)h(j+1)T - (j+1)(\rho_1 + \rho_2)\delta\},$$

And also, for $t \in [h(j+1)T + \delta(j+2)T]$, it follows from above results that

$$\|e(t)\|$$
$$\le \|e(0)\| \exp\{-\rho_1(h(j+1)T + \delta) + (\rho_1 + \rho_2)h(j+1)T$$
$$- (j+1)(\rho_1 + \rho_2)\delta\} \exp\{\rho_2(t - h(j+1)T - \delta)\}$$
$$= \|e(0)\| \exp\{\rho_2 t - (j+2)(\rho_1 + \rho_2)\delta\}.$$

From above discussion, we can see that the (14) is always correct for any nonnegative integer n.

When $h(n)$ is a strictly monotone increasing function on n and $h(n)T \le t < h(n)T + \delta$, it is easy to obtain

$$\frac{t-\delta}{T} < h(n) \le \frac{t}{T}, h^{-1}\left(\frac{t-\delta}{T}\right) < n \le h^{-1}\left(\frac{t}{T}\right),$$

$$-\rho_1 t + (\rho_1 + \rho_2) h(n) T - n(\rho_1 + \rho_2)\delta$$

$$\le -\rho_1 t + (\rho_1 + \rho_2) t - (\rho_1 + \rho_2)\delta h^{-1}\left(\frac{t-\delta}{T}\right)$$

$$= \rho_2 t - (\rho_1 + \rho_2)\delta h^{-1}\left(\frac{t-\delta}{T}\right)$$

$$= -\left[(\rho_1 + \rho_2)\delta\frac{h^{-1}\left(\dfrac{t-\delta}{T}\right)}{t} - \rho_2\right] t.$$

When $h(n)$ is a strictly monotone increasing function on n and $h(n)T + \delta \le t < h(n+1)T$, it follows that

$$h^{-1}\left(\frac{t}{T}\right) < (n+1) \le h^{-1}\left(\frac{t-\delta}{T}\right) + 1,$$

$$\rho_2 t - (n+1)(\rho_1 + \rho_2)\delta \le \rho_2 t - (\rho_1 + \rho_2) h^{-1}\left(\frac{t}{T}\right)$$

$$= -\left[(\rho_1 + \rho_2)\delta\frac{h^{-1}\left(\dfrac{t}{T}\right)}{t} - \rho_2\right] t,$$

$$h^{-1}\left(\frac{t-\delta}{T}\right) < h^{-1}\left(\frac{t}{T}\right),$$

Then

$$\rho_2 t - (n+1)(\rho_1 + \rho_2)\delta$$

$$\leq - \left[(\rho_1 + \rho_2)\delta \frac{h^{-1}\left(\dfrac{t-\delta}{T}\right)}{t} - \rho_2 \right] t.$$

Therefore

$$\left\| e(t) \right\| \leq \left\| e(0) \right\| \exp\left\{ -\left[(\rho_1 + \rho_2)\delta \frac{h^{-1}(t-\delta/T)}{t} - \rho_2 \right] t \right\}$$

$$\leq \left\| e(0) \right\| \exp\{-\zeta t\}, \quad t \in \left[h(n)T, h(n+1)T \right],$$

When $n \to +\infty, t \to \infty, \left\| e(t) \right\| \to 0$ is obtained under the condition (11). So the synchronization of the network (6) is achieved.

When $h(n)$ is a strictly monotone decreasing function on n with $\lim\limits_{n\to\infty} h(n) = 0, h(0) = +\infty$, we obtain the following result:

Theorem 3: Suppose that the operator g in the network (6) satisfies condition (5), and $\mu(-C)$ is defined as Definition 2, $\rho_1 = -\eta_1 - le^{-\eta_1 \tau}, \eta_1 = \mu(-C) + \lambda - k(1+\Delta), \rho_2 = \eta + le^{-\eta_2 \tau}, \eta_2 = \mu(-C) + \lambda,$ Where the constant

$$\lambda = \max_{\forall t \geq 0} \frac{\left(x(t) - s(t) \right)^{\mathrm{T}} \left(f\left(x(t), s(t) \right) - m\left(x(t), s(t) \right) \right)}{\left\| x(t) - s(t) \right\|^2},$$

e(t) Are the same as Theorem 2? So the synchronization of networks (6) is achieved if the parameters δ, T, k, λ, and ζ satisfy

$$P_1 > 0, \ P_2 > 0, \ \inf\left[(P_1 + P_2)\delta\frac{h^{-1}(t/T)}{t} - P_2 \right] \geq \zeta > 0. \tag{15}$$

Where $h^{-1}(\bullet)$ is the inverse function of the function? $h(\bullet)$.

Proof: From Theorem 1, the following conclusion is valid:

$$\|e(t)\| \leq \|e(h(n+1)T)\|\exp\{-P_1(t - h(n+1)T)\} \tag{16}$$

For any $h(n+1)T \leq t < h(n+1)T + \delta,$;

$$\|e(t)\| \leq \|e(h(n+1)T+\delta)\|\exp\{P_2(t - h(n+1)T - \delta)\} \tag{17}$$

For any $h(n+1)T + \delta \leq t < h(n)T$

From (16) and (17), imitating Theorem 2, we can prove

$$\|e(t)\| \leq \begin{cases} \|e(h(n+1)T)\|\exp\{g_1(t,n)\}, & (h(n+1)T \leq t < h(n+1)T+\delta), \\ \|e(h(n+1)T)\|\exp\{g_2(t,n)\}, & (h(n+1)T+\delta \leq t < h(n)T), \end{cases}$$

$$\leq \begin{cases} \|e(h(n+1)T)\|\exp\{g_3(t,n)\}, & (h(n+1)T \leq t < h(n+1)T+\delta), \\ \|e(h(n+1)T)\|\exp\{g_4(t,n)\}, & (h(n+1)T+\delta \leq t < h(n)T), \end{cases}$$

$$\leq \|e(0)\|\exp\left\{-\left[(P_1+P_2)\delta\frac{h^{-1}(t/T)}{t} - P_2\right]t\right\} \leq \|e(0)\|\exp\{-\zeta t\},$$

Where

$$g_1(t,n) = -\rho_1 t + (\rho_1 + \rho_2) h(n+1)T - (n+1)(\rho_1 + \rho_2)\delta,$$

$$g_2(t,n) = \rho_2 t - (n+2)(\rho_1 + \rho_2)\delta,$$

$$g_3(t,n) = -\left((\rho_1 + \rho_2)\delta \frac{h^{-1}(t/T)}{t} - \rho_2 \right) t,$$

$$g_4(t,n) = -\left((\rho_1 + \rho_2)\delta \frac{h^{-1}((t-\delta)/T)}{t} - \rho_2 \right) t.$$

When $t \to +\infty$, $\|e(t)\| \to 0$ is obtained under the condition (15). So the synchronization of network (6) is achieved.

Corollary 1: Supposing that $h(n) = p_1 n, \delta = p_2 T, p1 > 0, p_2 > 0$, and the rest of restricted conditions are invariable. Then the synchronization of the network (6) is achieved if the parameters δ, T and k, ζ satisfy

$$p_1 > 0, p_2 > 0 (p_1 + p_2)\delta \frac{p_2}{p_1} - p_2 \geq \zeta > 0$$

Corollary 2: when we add normally distributed white noise randn (size (t)), the result similar to Theorem 2 and Theorem 3 is obtained if the condition (11) or (12), respectively, is satisfied.

In the simulations of following examples, we always choose $N = 5, T = 4, \delta = 2.4, k = 10$ the matrix

$$A = \begin{pmatrix} -5 & 4 & 1 & 0 & 0 \\ 2 & -6 & 0 & 2 & 2 \\ 0 & 1 & -1 & 0 & 0 \\ 3 & 1 & 0 & -4 & 0 \\ 0 & 0 & 0 & 2 & -2 \end{pmatrix}.$$

Let the initial condition be

$$\left(x_1^T, x_2^T, x_3^T, x_4^T, x_5^T, s^T \right)$$
$$= \left(17, 12.8, 0.5, 0.6, 0.7, 0.8, 1, 1.3, 1.8, 1.9, 3, 4 \right).$$

Example 1 Consider a delayed system [47]:

$$\begin{cases} \dfrac{dx_1(t)}{dt} = -0.1x_1(t) + 0.4\sin x_2(t-2) \\[3mm] \dfrac{dx_2(t)}{dt} = -0.1x_2(t) + 0.3\sin x_1(t-2). \end{cases} \tag{18}$$

The function $h(n) = 2n = \ln(n+1), (n+1)/n^2, h(n) = 3/n+$, which are the strictly monotone increasing or decreasing function on n, respectively, then they can be clearly seen that the synchronization of network (6), which is composed of system (18), is realized in Figures 1-4 (Excited by parameter white-noise), respectively.

CONCLUSIONS

Approaches for quantized synchronization of complex networks with delayed time via general intermittent which uses the nonlinear operator named the matrix measure have been presented in this paper. Strong properties of global and exponential synchronization have been achieved in a finite number of steps. Numerical simulations have verified the effectiveness of the method.

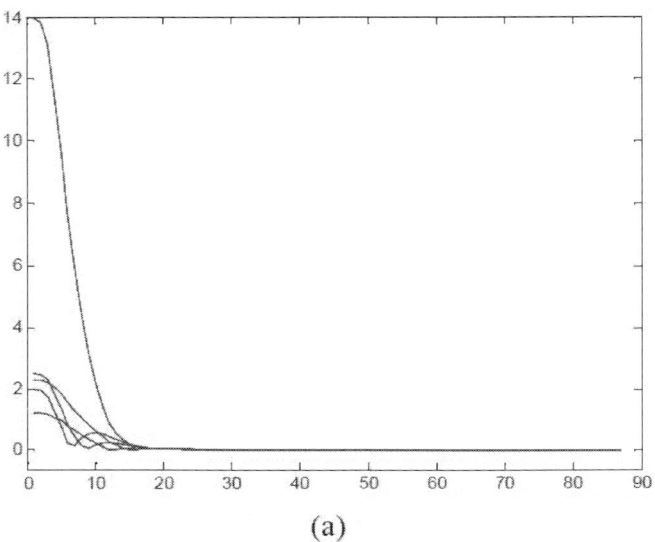

(a)

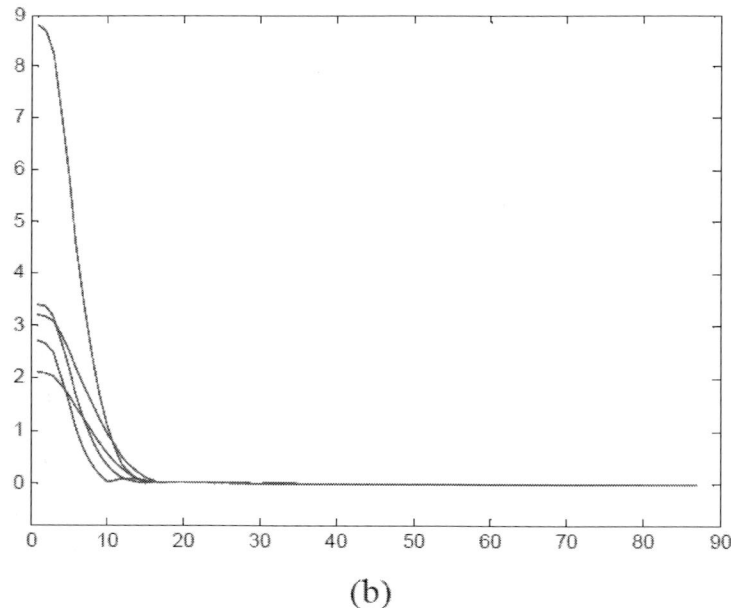

(b)

Figure 1: Synchronization error when h (n) = 2n + 1n (n + 1). (a) The error $x_{i1} - s_1$, (i = 1, 2, 3, 4, 5); (b) the error $x_{i2} - s_2$, (i = 1, 2, 3, 4, 5).

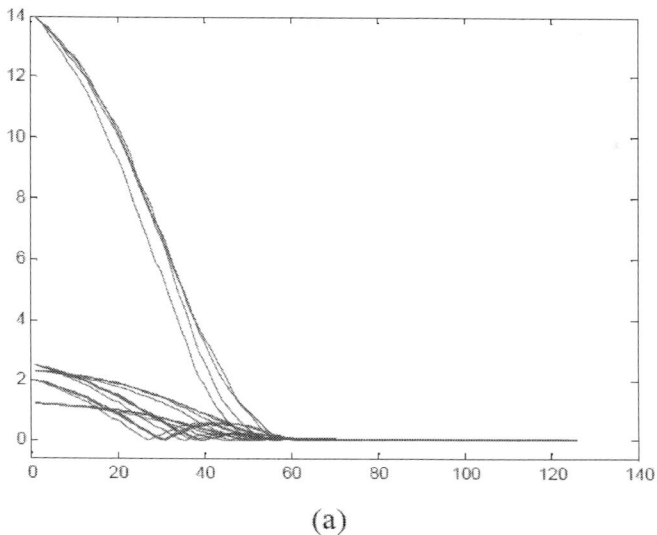

(a)

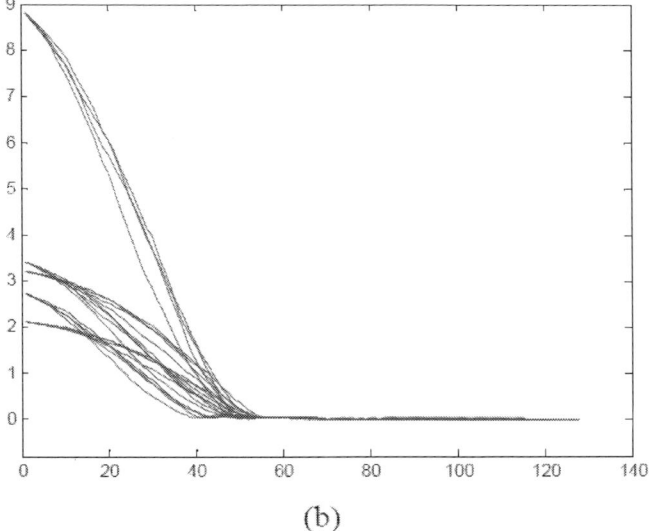

(b)

Figure 2: Synchronization error when h(n) = 2n + 1n(n + 1), white noise 0.5 (x$_i$ − s) randn (size(t)), (i = 1, 2, 3, 4, 5). (a) The error x$_{i1}$ − s$_1$, (i = 1, 2, 3, 4, 5); (b). The error x$_{i2}$ − s$_2$, (i = 1, 2, 3, 4, 5).

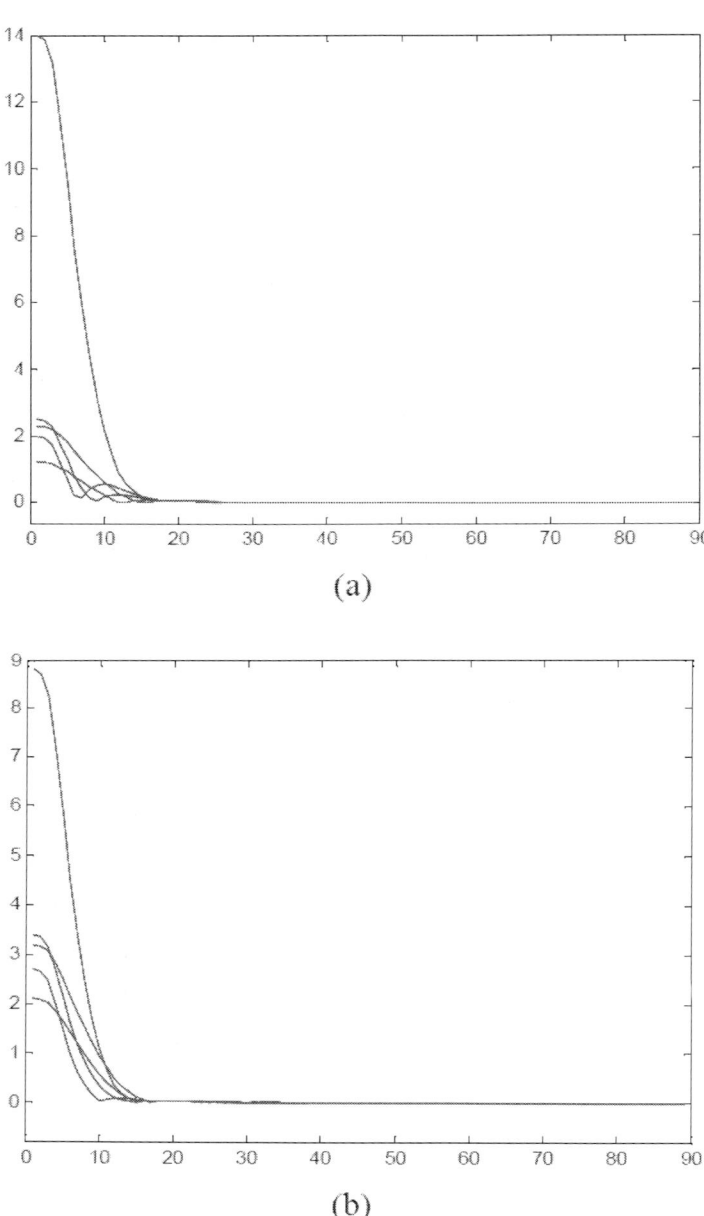

Figure 3: Synchronization error when h (n) = 3/n + (n + 1)/n². (a) The error $x_{i1} - s_1$, (i = 1, 2, 3, 4, 5); (b) the error $x_{i2} - s_2$, (i = 1, 2, 3, 4, 5).

Matrix Measure with Application in Quantized Synchronization

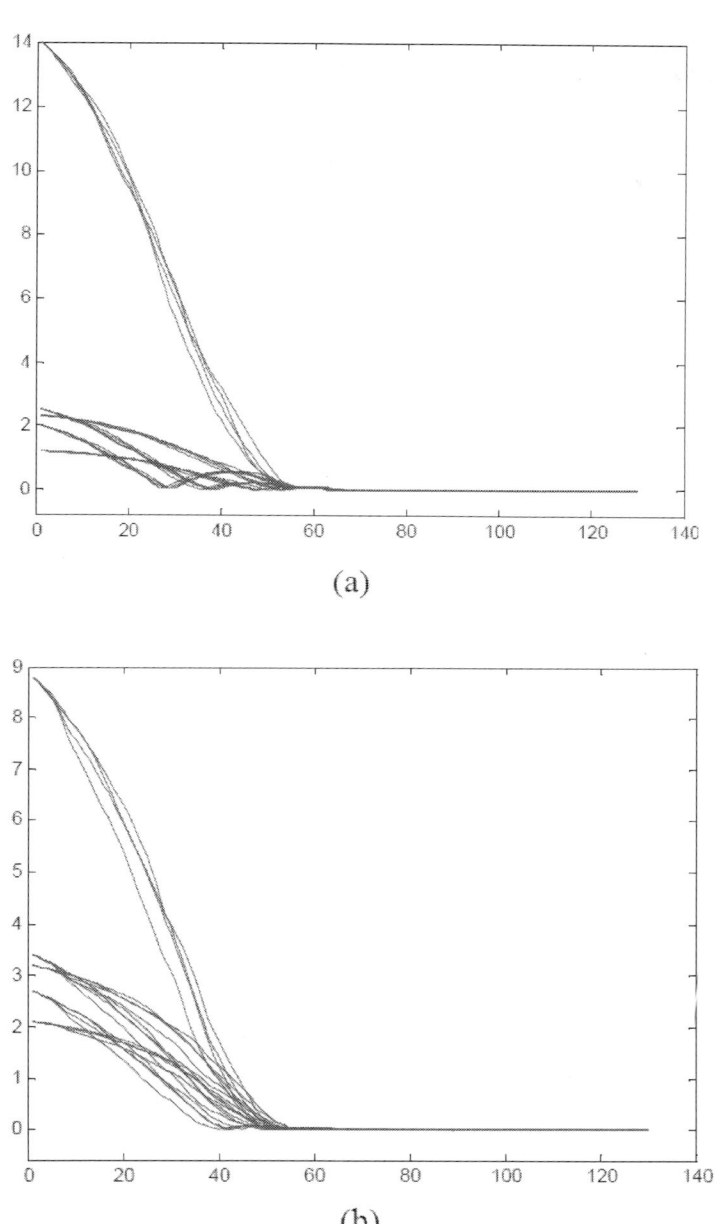

(a)

(b)

Figure 4: Synchronization error when $h(n) = 3/n + (n+1)/n^2$, white noise $0.5(x_i - s)$ randn (size (t)), ($i = 1, 2, 3, 4, 5$). (a) The error $x_{i1} - s_1$, ($i = 1, 2, 3, 4, 5$); (b) the error $x_{i2} - s_2$, ($i = 1, 2, 3, 4, 5$).

REFERENCES

1. H. Fujisaka and T. Yamada, "Stability Theory of Synchronized Motion in a Coupled-Oscillator System," Progress of Theoretical Physics, Vol. 69, No. 1, 1983, pp. 32-47. http://dx.doi.org/10.1143/PTP.69.32
2. H. Fujisaka and T. Yamada, "A New Intermittency in Coupled Dynamical Systems," Progress of Theoretical Physics, Vol. 74, No. 4, 1985, pp. 918-921. http://dx.doi.org/10.1143/PTP.74.918
3. H. Fujisaka and T. Yamada, "Stability Theory of Synchronized Motion in Coupled-Oscillator Systems IV," Progress of Theoretical Physics, Vol. 75, No. 5, 1986, pp. 1087-1104. http://dx.doi.org/10.1143/PTP.75.1087
4. V. S. Afraimovich, N. N. Verichev and M. I. Rabinovich, "Stochastic Synchronization of Oscillations in Dissipative Systems," Radiophysics and Quantum Electronics, Vol. 29, No. 9, 1986, pp. 747-751.
5. L. M. Pecora and T. L. Carroll, "Synchronization in Chaotic Systems," Physical Review Letters, Vol. 64, No. 8, 1990, pp. 821-824. http://dx.doi.org/10.1103/PhysRevLett.64.821
6. W. H. Deng, J. H. Lu and C.-P. Li, "Stability of N-Dimensional Linear Systems with Multiple Delays and Application to Synchronization," Journal of Systems Science and Complexity, Vol. 19, No. 2, 2006, pp. 149-156.
7. E. M. Elabbasy, H. N. Agiza and M. M. El-Dessoky, "Global Synchronization Criterion and Adaptive Synchronization for New Chaotic System," Chaos, Solitons and Fractals, Vol. 23, No. 4, 2005, pp. 1299-1309.
8. Q. L. Zhang, J. Zhou and G. Zhang, "Stability Concerning Partial Variables for a Class of Time-Varying Systems and Its Applications in Chaos Synchronization," Proceedings of the 24th Chinese Control Conference, South China University of Technology Press, Guangzhou, 2005, pp. 135-139.
9. Q. L. Zhang and G. J. Jia, "Chaos Synchronization of Morse Oscillator via Backstepping Design," Annals of Differential Equations, Vol. 22, No. 3, 2006, pp. 456-460.
10. Q. L. Zhang, "Synchronization of Multi-Chaotic Systems via Ring Impulsive Control," Control Theory & Applications, Vol. 27, No. 2, 2010, pp. 226-232.
11. Z. Tang and J. W. Feng, "Adaptive Cluster Synchronization for Nondelayed and Delayed Coupling Complex Networks with Nonidentical Nodes," Abstract and Applied Analysis, Vol. 2013, 2013, Article ID: 946243.
12. B. Li and Q. K. Song, "Synchronization of Chaotic Delayed Fuzzy Neural Networks under Impulsive and Stochastic Perturbations," Abstract and Applied Analysis, Vol. 2013, 2013, Article ID: 543549.
13. M. Han, Y. Liu and J. Q. Lu, "Impulsive Control for the Synchronization of Chaotic Systems with Time Delay," Abstract and Applied Analysis, Vol. 2013, 2013, Article ID: 647561.

14. S. J. Dan, S. X. Yang and W. Feng, "Lag Synchronization of Coupled Delayed Chaotic Neural Networks by Periodically Intermittent Control," Abstract and Applied Analysis, Vol. 2013, 2013, Article ID: 501461.

15. C.-M. Lin, M.-H. Lin and R.-G. Yeh, "Synchronization of Unified Chaotic System via Adaptive Wavelet Cerebellar Model Articulation Controller," Neural Computing and Applications, 2012, pp. 1-9. http://dx.doi.org/10.1007/s00521-012-1021-3

16. J. D. Cao and L. L. Li, "Cluster Synchronization in an Array of Hybrid Coupled Neural Networks with Delay," Neural Networks, Vol. 22, No. 4, 2009, pp. 335-342.

17. L. L. Li and J. D. Cao, "Cluster Synchronization in an Array of Coupled Stochastic Delayed Neural Networks via Pinning Control," Neurocomputing, Vol. 74, No. 5, 2011, pp. 846-856. http://dx.doi.org/10.1016/j.neucom.2010.11.006

18. H. Reza and P. Mass, "Delay-Range-Dependent Exponential H∞ Synchronization of a Class of Delayed Neural Networks," Chaos, Solitons and Fractals, Vol. 41, No. 3, 2009, pp. 1125-1135. http://dx.doi.org/10.1016/j.chaos.2008.04.051

19. J. D. Cao, D. W. C. Ho and Y. Q. Yang, "Projective Synchronization of a Class of Delayed Chaotic Systems via Impulsive Control," Physics Letters A, Vol. 373, No. 35, 2009, pp. 3128-3133. http://dx.doi.org/10.1016/j.physleta.2009.06.056

20. Z. Zahreddine, "Matrix Measure and Application to Stability of Matrices and Interval Dynamical Systems," International Journal of Mathematics and Mathematical Sciences, Vol. 2003, No. 2, 2003, pp. 75-85. http://dx.doi.org/10.1155/S0161171203202295

21. J. T. Sun, Y. P. Zhang, Y. Q. Liu and F. Q. Deng, "Exponential Stability of Interval Dynamical System with Multidelay," Applied Mathematics and Mechanics, Vol. 23, No. 1, 2002, pp. 87-91.

22. G. D. Zong, Y. Q. Wu and S. Y. Xu, "Stability Criteria for Switched Linear Systems with Time-Delay," Control Theory & Applications, Vol. 25, No. 5, 2008, pp. 295-305.

23. Y. H. Yuan, Q. L. Zhang and B. Chen, "Robust Fuzzy Control Based on Matrix Measure for Nonlinear Descriptor Systems with Time-Delay," Control and Decision, Vol. 22, No. 2, 2007, pp. 174-178.

24. L. M. Ding, J. B. Quan, Y. Zhang and Z. X. Li, "The Fixed Point Theorem and Asymptotic Stability of a Delay-Differential System," Journal of Air Force Radar Academy, Vol. 23, No. 3, 2009, pp. 203-204.

25. Y. X. Guo, "Matrix Measure and Uniform Ultimate Boundedness with Respect to Partial Variables for FDEs," Journal of Wuhan University of Science and Engineering, Vol. 21, No. 4, 2008, pp. 15-19.

26. J. J. Huang, C. D. Li and Q. Han, "Stabilization of Delayed Chaotic Neural Networks by Periodically Intermittent Control," Circuits, Systems & Signal Processing, Vol. 28, No. 4, 2009, pp. 567-579. http://dx.doi.org/10.1007/s00034-009-9098-3

27. J. Yu, H. J. Jiang and Z. D. Teng, "Synchronization of Nonlinear Systems with Delay via Periodically Intermittent Control," Journal of Xinjiang University (Natural Science Edition), Vol. 28, No. 3, 2010, pp. 310-315.

28. Z. Y. Dong, Y. J. Wang, M. H. Bai and Z. Q. Zuo, "Exponential Synchronization of Uncertain Master-Slave Lur'e Systems via Intermittent Control," Journal of Dynamics and Control, Vol. 7, No. 4, 2009, pp. 328-333.

29. W. G. Xia and J. D. Cao, "Pinning Synchronization of Delayed Dynamical Networks via Periodically Intermittent Control," Chaos, Vol. 19, No. 1, 2009, and Article ID: 013120. http://dx.doi.org/10.1063/1.3071933

30. P. Smolen, D. Baxter and J. Byrne, "Mathematical Modeling of Gene Networks Review," Neuron, Vol. 26, No. 3, 2000, pp. 567-580. http://dx.doi.org/10.1016/S0896-6273(00)81194-0

31. Y. Wang, J. Hao and Z. Zuo, "A New Method for Exponential Synchronization of Chaotic Delayed Systems via Intermittent Control," Physics Letters A, Vol. 374, No. 19-20, 2010, pp. 2024-2029. http://dx.doi.org/10.1016/j.physleta.2010.02.069

32. T. Huang, C. Li, W. Yu and G. Chen, "Synchronization of Delayed Chaotic Systems with Parameter Mismatches by Using Intermittent Linear State Feedback," Nonlinearity, Vol. 22, No. 3, 2009, pp. 569-584. http://dx.doi.org/10.1088/0951-7715/22/3/004

33. H. Zhang, H. C. Yan, F. W. Yang and Q. J. Chen, "Quantized Control Design for Impulsive Fuzzy Networked Systems," IEEE Transactions on Fuzzy Systems, Vol. 19, No. 6, 2011, pp. 1153-1162. http://dx.doi.org/10.1109/TFUZZ.2011.2162525

34. F. Fagnani and S. Zampieri, "Quantized Stabilization of Linear Systems: Complexity versus Performance," IEEE Transactions on Automatic Control, Vol. 49, No. 9, 2004, pp. 1534-1548. http://dx.doi.org/10.1109/TAC.2004.834111

35. S. Sqartini, B. Schuller and A. Hussain, "Cognitive and Emotionan Infromation Processing for Human-Machine Interaction," Cognitive Computation, Vol. 4, No. 4, 2012, pp. 383-387. http://dx.doi.org/10.1007/s12559-012-9180-1

36. J. B. Kim, J. S. Park and Y. H. Oh, "Speaker-Characterized Emotion Recognition Using Online and Iterative Speaker Adaptation," Cognitive Computation, Vol. 4, No. 4, 2012, pp. 398-408. http://dx.doi.org/10.1007/s12559-012-9132-9

37. L. I. Per lovsky and D. S. Levine, "The Drive for Creativity and the Escape from Creativity: Neurocognitive Mechanisms," Cognitive Computation, Vol. 4, No. 3, 2012, pp. 292-305.

38. Q. L. Zhang, "The Generalized Dahlquist Constant with Applications in Synchronization Analysis of Typical Neural Networks via General Intermittent Control," Advances in Artificial Neural Systems, Vol. 2011, 2011, Article ID: 249136. http://dx.doi.org/10.1155/2011/249136

39. G. T. Hui, B. N. Huang, Y. C. Wang and X. P. Meng, "Quantized Control Design for Coupled Dynamic Networks with Communication Constraints," Cognitive Computation, Vol. 5, No. 2, 2013, pp. 200-206.

40. L. A. Montestruque and P. J. Antsaklis, "Static and Dynamic Quantization in Model-Based Networked Control Systems," International Journal of Control, Vol.

80, No. 1, 2007, pp. 87-101. http://dx.doi.org/10.1080/00207170600931663

41. Q. Ye, H. B. Zhu and B. T. Cui, "Synchronization Analysis of Delayed Hybrid Dynamical Networks with Quantized Impulsive Effects," Control Theory & Applications, Vol. 30, No. 1, 2013, pp. 61-68.

42. W. Liu, Z. M. Wang and M. K. Ni, "Quantized Feedback Stabilization of Model-Based Networked Control Systems," Control and Decision, Vol. 28, No. 2, 2013, pp. 285-288.

43. R. Q. Lu, F. Wu and A. K. Xue, "A Reset Quantized-State Controller for Linear Systems," Control Theory & Applications, Vol. 29, No. 4, 2012, pp. 507-512.

44. X. X. Liao, "Theory and Application of Stability for Dynamical Systems," National Defence Industry Press, Beijing, 2000, pp. 15-40.

45. B. Shi, D. C. Zhang and M. J. Gai, "Theory and Applications of Differential Equations," National Defense Industry Press, Beijing, 2005, pp. 18-23.

46. J. C. Kuang, "Applied Inequalities," Shandong Science and Technology Press, Jinan, 2004, pp. 564-570.

47. C. Liu, C. D. Li and S. K. Duan, "Stabilization of Oscillating Neural Networks with Time-Delay by Intermittent Control," International Journal of Control, Automation and Systems, Vol. 9, No. 6, 2011, pp. 1074-1079. http://dx.doi.org/10.1007/s12555-011-0607-3

CITATION

Zhang, Q. (2013) Matrix Measure with Application in Quantized Synchronization Analysis of Complex Networks with Delayed Time via the General Intermittent Control. Applied Mathematics, 4, 1417-1426 doi: 10.4236/am.2013.410192.

On Humbert Matrix Polynomials of Two Variables

Ghazi S. Khammash[1] and A. Shehata[2]

[1]Department of Mathematics, Al-Aqsa University,
Gaza Strip, Palestine
[2]Department of Mathematics, Faculty of
Science, Assiut University, Assiut, Egypt

ABSTRACT

In this paper we introduce Humbert matrix polynomials of two variables. Some hypergeometric matrix representations of the Humbert matrix polynomials of two variables, the double generating matrix functions and expansions of the Humbert matrix polynomials of two variables in series of Hermite polynomials are given. Results of Gegenbauer matrix polynomials of two variables follow as particular cases of Humbert matrix polynomials of two variables.

INTRODUCTION

The special matrix functions appear in statistics, lie group theory and number theory [1-4] and the matrix polynomials have become more important and some results in the theory of classical orthogonal polynomials have been extended to orthogonal matrix polynomials see for instance [5-9].

If D_0 is the complex plane cut along the negative, real axis and log (z) denotes the principal logarithm of z (Saks, S. and A. Zygmund, [10]), then $z^{1/2}$ represents exp(1/2log)(z) if A is a matrix in $C^{N \times N}$ the set of all the eigenvalues of is denoted by the set of all the eigenvalues of A is denoted by $\sigma(A)$. If f(z) and g(z) are holomorphic functions of the complex variable z, which are defined in an open set Ω of the complex plane, and A is a matrix in $C^{N \times N}$ such that $\sigma(A) \subset \Omega$. Then from the properties of the matrix functional calculus, (Dunford N. and J. Schwartz J. [11]), it follows that $f(A)g(A)=g(A)f(A)$. If A is a matrix with $\sigma(A) \subset D_0$, then $A^{1/2} = \sqrt{A}$ denotes the a image by $z^{1/2}$ of the matrix functional calculus acting on the matrix A. we say that A is a positive stable matrix if $Re(z)>0$ for all $z \in \sigma(A)$.

For any matrix P in $C^{N \times N}$ we will exploit the following relation due to [12]

$$(1-x)^{-p} = \sum_{n=0}^{\infty} \frac{(p)_n x^n}{n!}, \quad |x|<1 \tag{1}$$

Khammash [12], define the Gegenbauer matrix polynomials of two variables by

$$C_{n,k}^A (x,y) = \sum_{r=0}^{\left[\frac{n}{2}\right]} \sum_{j=0}^{\left[\frac{k}{2}\right]} \frac{(-1)^{r+j} (A)_{n+k-r-j} (2x)^{n-2r} (2y)^{k-2j}}{r! j! (n-2r)! (k-2j)!} \tag{2}$$

From which it follows that $C_{n,k}^A(x,y)$ is a matrix polynomial in two variables x and y of degree precisely n in x and k in y.

Also we recall that if A(k, n) are matrix in $C^{N \times N}$ for $n \geq 0$ and $k \geq 0$ that it follows that (Defez and Jódar [14])

$$\sum_{n=0}^{\infty} \sum_{k=0}^{\infty} A(k,n) = \sum_{n=0}^{\infty} \sum_{k=0}^{n} A(k,n-k) \tag{3}$$

$$\sum_{n=0}^{\infty} \sum_{k=0}^{\infty} A(k,n) = \sum_{n=0}^{\infty} \sum_{k=0}^{[n/2]} A(k,n-k) \tag{4}$$

and, for m is a positive integer such that n>m, then

$$\sum_{n=0}^{\infty} \sum_{k=0}^{\infty} A(k,n) = \sum_{n=0}^{\infty} \sum_{k=0}^{[n/m]} A(k,n-k) \tag{5}$$

$$\sum_{n=0k=0}^{\infty \ n} A(k,n) = \sum_{n=0 \ k=0}^{\infty \ [n/m]} A(k,n-mk+k) \qquad (6)$$

We define Humbert matrix polynomials of two variables and discuss its special cases. Some hypergeometric matrix representations of the Humbert matrix polynomials of two variables, the double generating matrix functions and expansions of the Humbert matrix polynomials of two variables in series of Hermite polynomials are given. Some particular cases are also discussed.

DEFINITION OF HUMBERT MATRIX POLYNOMIALS OF TWO VARIABLES

Let A be a positive stable matrix in $C^{N \times N}$ for a positive integer m, we define Humbert matrix polynomials by

$$\left(1-\left(mxt - t^{m}\right) - \left(mys - s^{m}\right)\right)^{-A}$$

$$= \sum_{n=0k=0}^{\infty \ \infty} P_{n,k,m}\left(x,y,A\right)t^{n}s^{k} \qquad (7)$$

Now (7) it can be written in the form

$$\left(1-\left(mxt - t^{m}\right) - \left(mys - s^{m}\right)\right)^{-A}$$

$$= \sum_{n=0k=0}^{\infty \ \infty} \frac{\left(A\right)_{n+k}\left(mxt - t^{m}\right)^{n}\left(mys - s^{m}\right)^{k}}{n!k!}$$

and by (3) and (6) respectively, one gets

$$= \sum_{n=0k=0r=0j=0}^{\infty \ \infty \ n \ k} \frac{\left(A\right)_{n+k}\left(mx\right)^{n-r}\left(my\right)^{k-j}\left(-1\right)^{r+j}}{r!j!(n-r)!(k-j)!}$$

$$\times t^{n+(m-1)r}s^{k+(m-1)j}$$

$$= \sum_{n=0}^{\infty} \sum_{k=0}^{\infty} \sum_{r=0}^{[n/m]} \sum_{j=0}^{[k/m]} \frac{(A)_{(n+k)-(m-l)(r+j)} (-1)^{r+j}}{r! \, j! \, (n-mr)! \, (k-mj)!}$$

$$\times (mx)^{n-mr} (my)^{k-mj} t^n s^k$$

$$\sum_{n=0}^{\infty} \sum_{k=0}^{\infty} P_{n,k,m} (x, y, A) s^n t^k$$

$$= \sum_{n=0}^{\infty} \sum_{k=0}^{\infty} \sum_{r=0}^{[n/m]} \sum_{j=0}^{[k/m]} \frac{(A)_{(n+k)-(m-l)(r+j)} (-1)^{r+j}}{r! \, j! \, (n-mr)! \, (k-mj)!}$$

$$\times (mx)^{n-r} (my)^{k-j} t^n s^k \tag{8}$$

By equating the coefficients of $t^n s^k$ in (7) and (8), we obtain an explicit representation of the Humbert matrix polynomials of two variables. In the form

$$= \sum_{r=0}^{[n/m]} \sum_{j=0}^{[k/m]} \frac{(A)_{(n+k)-(m-l)(r+j)} (-1)^{r+j} \times (mx)^{n-mr} (my)^{k-mj}}{r! \, j! \, (n-mr)! \, (k-mj)!} \tag{9}$$

from which it follows that $P_{n'k'm}(x,y,A)$ is a matrix polynomial in two variables x and y of degree precisely n in x and k in y. In (9) setting m = 2, we get the Gegenbauer matrix polynomials of two variables [13] as particular case of the Humbert matrix polynomials of two variables.

HYPERGEOMETRIC MATRIX REPRESENTATION FOR $P_{n,k,m}(x,y,A)$

We study here the representation of the hypergeometric matrix representation for the Humbert matrix polynomials of two variables. There are some facts and notations used throughout the development in Sections 3 - 5, which are listed here.

Fact 1 [15]: The reciprocal scalar Gamma Function $\Gamma^{-1}(z) = 1/\Gamma(z)$, is an entire functions of the complex variable z. Thus, for $c \in C^{N \times N}$, the Riesz-Dunford functional calculus [11] shows that $\Gamma^{-1}(z)$ is well defined and is indeed, the inverse of $\Gamma(z)$, Hence: if $c \in C^{r \times r}$ is such that C+nI is invertible for every integer n≥0. Then

$$(c)_n = \Gamma(c+nI)\Gamma^{-1}(z).$$

Fact 2 [12]: If A, B and C are members of $C^{r \times r}$ for which C+nI is invertible for every integer n≥0. The hypergeometric matrix function F(A, B;C;z) is defined by

$$F(A,B;C;z) = \sum_{n=0}^{\infty} \frac{1}{n!}\left((A)_n (B)_n \left[(C)_n\right]^{-1}\right) z^n$$

it converges for $|z| < 1$.

Notation 1 [16]: For $A \in C^{r \times r}$, the matrix version of the pochhammer symbol (the shifted factorial) is

$$(A)_n = A(A+I)(A+2I)\cdots(A+(n-1)I),$$
$$n \geq 1 \tag{10}$$

With $(A)_0 = I$.

Note that A=-jI, where j is a positive integer, then $(A)_n = 0$ when ever n>j. Also, the product in (10) is commutative, and then it is easy to see that

$$(A)_{n+k} (A)_n (A+nI)_k$$

$$(A)_{n-k} = (-1)^k (A)_n \left[(I-A-nI)\right]^{-k-1}$$

and

$$(A)_{mn} = m^{mn} \prod_{s=1}^{m}\left(\frac{1}{m}(A+(s-1))I\right)_n$$

where m is a positive integer.

Notation 2 [17]

$$\frac{1}{(n-mk)} = \frac{(-1)^{mk}}{n!}(-n)_m k; \quad 0 \le mk \le n$$

$$(-nI)_{mk} = m^{mk}\prod_{s=1}^{m}\left(\frac{1}{m}(s-n-1)I\right)_k$$

Now, in view of Notation 2, the explicit representation (9) for m≥2, becomes

$$P_{n,k,m}(x,y,A) = \sum_{r=0}^{\left[\frac{n}{m}\right]}\sum_{j=0}^{\left[\frac{k}{m}\right]}\frac{(A)_{(n+k)-(m-l)(r+j)}}{r!j!n!k!} \times (-1)^{(m+l)(r+j)}(mx)^{n-mr}(my)^{k-mj}$$

$$\times m^{mr}\prod_{s=1}^{m}\left(\frac{1}{m}(s-n-1)I\right)_r m^{mj}\prod_{l=1}^{m}\left(\frac{1}{m}(l-k-1)I\right)_j$$

$$= \frac{(A)_{n+k}(mx)^n(my)^k}{n!k!} \times \sum_{r=0}^{\left[\frac{n}{m}\right]}\sum_{j=0}^{\left[\frac{k}{m}\right]}\frac{(1)}{r!j!}\left[\prod_{s=1}^{m-1}\left(\frac{(-A-(n+k)I+sI)}{(m-1)}\right)_{r+j}\right]^{-1}$$

$$\times \prod_{s=1}^{m}\left(\frac{s-n-1}{m}I\right)_r\prod_{l=1}^{m}\left(\frac{l-k-1}{m}I\right)_j \times \left(\frac{1}{(m-1)^{m-1}x^m}\right)^r\left(\frac{1}{(m-1)^{m-1}y^m}\right)^j$$

Thus we get the following hypergeometric representation of Humbert matrix polynomials of two variables.

$$P_{n,k,m}(x,y,A) = \frac{(A)_{n+k}(mx)^n(my)^k}{n!k!} \times {}_mF_{m-1}^{(2)}\left[\frac{-n}{m}I, \frac{n-1}{m}I, \cdots, \frac{n-m+1}{m}I; \frac{-k}{m}I, \frac{k-1}{m}I, \cdots, \frac{k-m+1}{m}I; \right.$$

$$\left. \frac{-A-((n+k)-1)I}{m-1}, \frac{-A-((n+k)-2)I}{m-1}, \cdots, \frac{-A-((n+k)-(m-1))I}{m-1}; \frac{1}{(m-1)^{m-1}x^m}, \frac{1}{(m-1)^{m-1}y^m}\right]$$

$$(11)$$

For m = 2 (11), we gives hypergeometric representation of Gegenbauer matrix polynomials of two variables [13].

The above facts and notations will be used throughout the next two sections.

ADDITIONAL DOUBLE GENERATING MATRIX FUNCTIONS

Now, since

$$P_{n,k,m}(x,y,A) = \sum_{r=0}^{\left[\frac{n}{m}\right]} \sum_{j=0}^{\left[\frac{k}{m}\right]} \frac{(A)_{(n+k)-(m-l)(r+j)}}{r!\,j!\,(n-mr)!\,(k-mj)!} \times (-1)^{r+j} (mx)^{n-r} (my)^{k-j} \sum_{n=0}^{\infty}\sum_{k=0}^{\infty} \frac{P_{n,k,m}(x,y,A)\,s^n t^k}{(A)_{n+k}}$$

$$= \sum_{n=0}^{\infty}\sum_{k=0}^{\infty}\sum_{r=0}^{\left[\frac{n}{m}\right]}\sum_{j=0}^{\left[\frac{k}{m}\right]} \frac{(A)_{(n+k)-(m-l)(r+j)}}{(A)_{n+k}\,r!\,j!\,(n-mr)!\,(k-mj)!} \times (-1)^{r+j} (mx)^{n-r} (my)^{k-j}\, t^{k+mj} s^{n+mr}$$

$$\tag{12}$$

By using (5), one gets

$$= \sum_{n=0}^{\infty}\sum_{k=0}^{\infty}\sum_{r=0}^{\infty}\sum_{j=0}^{\infty} \frac{(A)_{(n+k)+(r+j)}}{(A)_{(n+mr)+(k+mj)}\,r!\,j!\,n!\,k!} \times (-1)^{r+j} (mx)^{n-r} (my)^{k-j}\, t^{k+mj} s^{n+mr}$$

$$\tag{13}$$

By using Notation 1, the following generating matrix functions for Humbert matrix polynomials of two variables follows

$$= \sum_{n=0}^{\infty}\sum_{k=0}^{\infty}\sum_{r=0}^{\infty}\sum_{j=0}^{\infty} \frac{\left(A+(n+k)I\right)_{(r+j)}}{\left(A+(n+k)I\right)_{m(r+j)}\,r!\,j!\,n!\,k!} \times (-1)^{r+j} (mx)^{n-r} (my)^{k-j}\, t^{k+mj} s^{n+mr}$$

$$= \sum_{n=0}^{\infty}\sum_{k=0}^{\infty}\sum_{r=0}^{\infty}\sum_{j=0}^{\infty} \frac{\left(A+(n+k)I\right)_{(r+j)}}{r!\,j!\,n!\,k!\,m^m \prod_{s=1}^{m}\left(\frac{1}{m}\left(A_{(n+k)}+(s+1)I\right)\right)_{(r+j)}} \times (-1)^{r+j} (mx)^{n-r} (my)^{k-j}\, t^{k+mj} s^{n+mr}$$

$$\sum_{n=0}^{\infty}\sum_{k=0}^{\infty} P_{n,k,m}(x,y,A)\left[(A)_{n+k}\right]^{-1} s^n t^k$$

$$= \sum_{n=0}^{\infty}\sum_{k=0}^{\infty} \frac{(mxt)^n (mys)^k}{n!k!} \sum_{r=0}^{\infty}\sum_{j=0}^{\infty} \frac{\left(A+(n+k)I\right)_{(r+j)} t^{k+mj} s^{n+mr}}{\prod_{s=1}^{m}\left(\frac{1}{m}\left(A_{(n+k)}+(s+1)I\right)\right)_{(r+j)} r!j!} \times \left(\left(\frac{-s}{m}\right)^m\right)^r\left(\left(\frac{-t}{m}\right)^m\right)^j$$

have thus discovered the family of double generating function of the Humbert polynomials of two variables

$$\sum_{n=0}^{\infty}\sum_{k=0}^{\infty} P_{n,k,m}(x,y,A)\left[(A)_{n+k}\right]^{-1} s^n t^k = \sum_{n=0}^{\infty}\sum_{k=0}^{\infty} \frac{(mxt)^n (mys)^k}{n!k!}$$

$$\times {}_1F_m^{(2)}\left[\left(A+(n+k)I\right):\frac{A+(n+k)I}{m}I,\frac{A+((n+k)+1)I}{m}I,\cdots,\frac{A+((n+k)+(m-1))I}{m}I:\left(\frac{-s}{m}\right)^m,\left(\frac{-t}{m}\right)^m\right]$$

(14)

If B is a positive stable matrix in $C^{N\times N}$, then let us now return to (12) and consider the double sum.

$$\sum_{n=0}^{\infty}\sum_{k=0}^{\infty} \frac{(B)_{n+k} P_{n,k,m}(x,y,A) t^k s^n}{(A)_{n+k}} = \sum_{n=0}^{\infty}\sum_{k=0}^{\infty}\sum_{r=0}^{\left[\frac{n}{m}\right]}\sum_{j=0}^{\left[\frac{k}{m}\right]} \frac{(B)_{n+k} (A)_{(n+k)-(m-i)(r+j)}}{(A)_{n+k} r!j!(n-mr)!(k-mj)!} \times (-1)^{r+j} (mx)^{n-r} (my)^{k-j} t^k s^n$$

Then in similar manner, we get

$$\sum_{n=0}^{\infty}\sum_{k=0}^{\infty} (B)_{n+k} P_{n,k,m}(x,y,A)\left[(A)_{n+k}\right]^{-1} t^k s^n$$

$$= \sum_{n=0}^{\infty}\sum_{k=0}^{\infty} \frac{(mxt)^n (mys)^k}{n!k!} \times {}_mF_{m-1}^{(2)}\left[\frac{B+(n+k)I}{m};\frac{B+((n+k)+1)I}{m},\frac{B+((n+k)+(m-1))I}{m},\right.$$

$$\left. A+(n+k)I;\frac{A+(n+k)I}{m};\frac{A+((n+k)+1)I}{m},\frac{A+((n+k)+(m-1))I}{m};-s^m,-t^m\right]$$

(15)

EXPANSIONS OF $P_{n,k,m}(x,y,A)$, IN SERIES OF HERMITE $H_{n,k}(x, y, A)$

Now, we derive expansions of $P_{n,k,m}(x, y, A)$ in series of Hermite $H_{n,k}(x, y, A)$ According to [18], the expansion of $x^n I$ in a series of Hermite matrix polynomials was given in the form:

$$\frac{x^n}{n!}I=\left(\sqrt{2A}\right)^{-n}\sum_{k=0}^{[n/2]}\frac{y^k}{k!(n-2k)!}H_{n-2k}(x,y,A)$$

(16)

which with the aid of (5) and (9), one gets

$$\sum_{n=0}^{\infty}\sum_{k=0}^{\infty}P_{n,k,m}(x,y,A)t^k s^n=\sum_{n=0}^{\infty}\sum_{k=0}^{\infty}\sum_{r=0}^{\left[\frac{n}{m}\right]}\sum_{j=0}^{\left[\frac{k}{m}\right]}\frac{(A)_{(n+k)-(m-l)(r+j)}}{r!j!(n-mr)!(k-mj)!}\times(-1)^{r+j}(mx)^{n-mr}(my)^{k-mj}t^k s^n$$

$$=\sum_{n=0}^{\infty}\sum_{k=0}^{\infty}\sum_{r=0}^{\infty}\sum_{j=0}^{\infty}\frac{(-1)^{r+j}(A)_{(n+k)(r+j)}}{r!j!n!k!}\times(mx)^n(my)^k t^{k+mj}s^{n+mr}$$

From (16), we get

$$=\sum_{n=0}^{\infty}\sum_{k=0}^{\infty}\sum_{r=0}^{\infty}\sum_{j=0}^{\infty}\sum_{p=0}^{[n/2]}\frac{y^p\left(\sqrt{2A}\right)^{-n}(A)_{(n+k)(r+j)}}{r!j!k!p!(n-2p)!}\times(-1)^{r+j}(my)^k H_{n-2p}(mx,y,A)s^{n+mr}t^{k+mj}$$

$$=\sum_{n=0}^{\infty}\sum_{k=0}^{\infty}\sum_{r=0}^{\infty}\sum_{j=0}^{\infty}\sum_{p=0}^{\infty}\frac{y^p\left(\sqrt{2A}\right)^{-n-2p}(A)_{((n+2p)+k)+(r+j)}}{r!j!k!p!n!}\times(-1)^{r+j}(my)^k H_n(mx,y,A)s^{n+2p+mr}t^{k+mj}$$

$$=\sum_{n=0}^{\infty}\sum_{k=0}^{\infty}\sum_{r=0}^{\left[\frac{n}{m}\right]}\sum_{j=0}^{\left[\frac{k}{m}\right]}\sum_{p=0}^{\min(r,j)}\frac{(-1)^{(r-p)+(j-p)}y^p}{(r-p)!(j-p)!}\times\frac{\left(\sqrt{2A}\right)^{-n-2m-2p}(my)^{k-mj}}{(k-mj)!(n-mr)!p!}\times(A)_{(n+k)-(m-1)(r+j)}H_{n-mr}(mx,y,A)s^{n+(2-m)p}t^k$$

Now replacing n by n+(m-2)P and equating the coefficients of $s^n t^k$, we get

$$P_{n,k,m}(x,y,A) = \sum_{r=0}^{\left[n+(m-2)p/m\right]} \sum_{j=0}^{\left[\frac{k}{m}\right]} \sum_{p=0}^{\min(r,j)} \frac{(-r)_p (-j)_p}{r!\,j!\,p!\,(k-mj)!} \times \frac{(-1)^{(r+j)} (m)^{k-mj} (y)^{k+p-mj}}{(n-2p-m(p+r))!}$$

$$\times \left(\sqrt{2A}\right)^{-n-m(p+r)} (A)_{(n+k)+(m-2p)-(m-l)(r+j)} \times H_{n-2p-m(r+p)}(mx,y,A)$$

(17)

REFERENCES

1. A. G. Constantine and R. J. Mairhead, "Partial Differential Equations for Hypergeometric Functions of Two Argument Matrices," Journal of Multivariate Analysis, Vol. 2, No. 3, 1972, pp. 332-338. doi:10.1016/0047-259X(72)90020-6.

2. R. J. Muirhead, "Systems of Partial Differential Equations for Hypergeometric Functions of Matrix Argument," Annals of Mathematical Statistics, Vol. 40, No. 3, 1970, pp. 991-1001. doi:10.1214/aoms/1177696975.

3. A. Terras, "Special Functions the Symmetric Space of Positive Matrices," SIAM Journal on Mathematical Analysis, Vol. 16, No. 3, 1985, pp. 620-640. doi:10.1137/0516046.

4. A. T. James, "Special Functions of Matrix and Single Argument in Statistics in Theory and Application of Special Functions," Academic Press, New York, 1975.

5. A. J. Duran, "Markov's Theorem for Orthogonal Matrix Polynomials," Canadian Journal of Mathematics, Vol. 48, 1996, pp. 1180-1195. doi:10.4153/CJM-1996-062-4.

6. A. J. Duran and W. Van Assche, "Orthogonal Matrix Polynomials and Higher Order Recurrence Relations," Linear Algebra and Its Applications, Vol. 219, No. 1, 1995, pp. 261-280. doi:10.1016/0024-3795(93)00218-O.

7. L. Jódar and J. Sastre, "The Growth of Laguerre Matrix Polynomials on Bounded Inervals," Applied Mathematics Letters, Vol. 13, No. 8, 2000, pp. 21- 26. doi:10.1016/S0893-9659(00)00090-2.

8. L. Jódar, R. Company and E. Navarro, "Laguerre Matrix Polynomials and Systems of Second Order Differential Equations," Applied Numerical Mathematics, Vol. 15, No. 1, 1994, pp. 53-63. doi:10.1016/0168-9274(94)00012-3.

9. A. Sinap and W. Van Assch, "Orthogonal Matrix Polynomials and Applications," Journal of Computational and Applied Mathematics, Vol. 66, No. 1-2, 1996, pp. 27-52. doi:10.1016/0377-0427(95)00193-X.

10. S. Saks and A. Zygmund, "Analytic Functions," Elsevier, Amsterdam, 1971.

11. N. Dunford and J. Schwartz, "Linear Operators, Part I," Interscience, New York, 1955.

12. L. Jódar and J. C. Cortés, "On the Hypergeometric Matrix Function," Journal of Computational and Applied Mathematics, Vol. 99, No. 1-2, 1998, pp. 205-217. doi:10.1016/S0377-0427(98)00158-7.
13. G. S. Khammash, "A Study of a Two Variables Gegenbauer Matrix Polynomials and Second Order Partial Matrix Differential Equations," International Journal of Mathematical Analysis, Vol. 2, No. 17, 2008, pp. 807-821.
14. E. Defez and L. Jódar, "Some Applications of the Hermite Matrix Polynomials Series Expansions," Journal of Computational and Applied Mathematics, Vol. 99, No. 1-2, 1998, pp. 105-117. doi:10.1016/S0377-0427(98)00149-6.
15. E. Hille, "Lectures on Ordinary Differential Equations," Addison-Wesley, New York, 1969.
16. L. Jódar and J. C. Cortés, "Some Properties of Gamma and Beta Matrix Functions," Applied Mathematics Letters, Vol. 11, No. 1, 1998, pp. 89-93. doi:10.1016/S0893-9659(97)00139-0.
17. E. D. Rainvlle, "Special Functions," Chelsea Publishing Co., Bronx, New York, 1960.
18. R. S. Batahan, "Anew Extension of Hermite Matrix Polynomials and Its Applications," Linear Algebra and Its Applications, Vol. 419, No. 1, 2006, pp. 82-92. doi:10.1016/j.laa.2006.04.006.

CITATION

G. Khammash and A. Shehata, "On Humbert Matrix Polynomials of Two Variables," Advances in Pure Mathematics, Vol. 2 No. 6, 2012, pp. 423-427. doi: 10.4236/apm.2012.26064.

Approximate Solution of Fuzzy Matrix Equations with LR Fuzzy Numbers

Xiaobin Guo[1] and Dequan Shang[2]

[1]College of Mathematics and Statistics, Northwest Normal University, Lanzhou, China
[2]Department of Public Courses, Gansu College of Chinese Medicine, Lanzhou, China

ABSTRACT

In the paper, a class of fuzzy matrix equations $A\tilde{X} = \tilde{B}$ where A is an m × n crisp matrix and $\tilde{B}$ is an m × p arbitrary LR fuzzy numbers matrix, is investigated. We convert the fuzzy matrix equation into two crisp matrix equations. Then the fuzzy approximate solution of the fuzzy matrix equation is obtained by solving two crisp matrix equations. The existence condition of the strong LR fuzzy solution to the fuzzy matrix equation is also discussed. Some examples are given to illustrate the proposed method. Our results enrich the fuzzy linear systems theory.

INTRODUCTION

Systems of simultaneous matrix equations are essential mathematical tools in science and technology. In many applications, at least some of the parameters of the system are represented by fuzzy rather than crisp numbers. So, it is very important to develop a numerical procedure that would appropriately handle and solve fuzzy matrix systems. The concept of fuzzy numbers and arithmetic operations were first introduced and investigated by Zadeh [1] and Dubois [2].

Since M. Friedman et al. [3] proposed a general model for solving a n×n fuzzy linear systems whose coefficients matrix is crisp and the right-hand side is a fuzzy number vector in 1998, many works have been done about how to deal with some advanced fuzzy linear systems such as dual fuzzy linear systems (DFLS), general fuzzy linear systems (GFLS), fully fuzzy linear systems (FFLS), dual fully fuzzy linear systems (DFFLS) and general dual fuzzy linear systems (GDFLS), see [4-9]. However, for a fuzzy linear matrix equation which always has a wide use in control theory and control engineering, few works have been done in the past decades. In 2010, Gong Zt [10,11]investigated a class of fuzzy matrix equations $A\tilde{X} = \tilde{B}$ by means of the undetermined coefficients method, and studied least squares solutions of the inconsistent fuzzy matrix equation by using generalized inverses. In 2011, Guo X. B. [12] studied the minimal fuzzy solution of fuzzy Sylvester matrix equations $A\tilde{X} + \tilde{X}B = \tilde{C}$. Recently, they [13] considered the fuzzy symmetric solutions of fuzzy matrix equations $A\tilde{X} = \tilde{B}$.

The LR fuzzy number and its operations were firstly introduced by Dubois [2]. In 2006, Dehgham et al. [6] discussed the computational methods for fully fuzzy linear systems whose coefficient matrix and the right-hand side vector are denoted by LR fuzzy numbers. In this paper, we propose a practical method for solving a class of fuzzy matrix system $A\tilde{X} = \tilde{B}$ in which A is an m × n crisp matrix and $\tilde{B}$ is an m × p arbitrary LR fuzzy numbers matrix. In contrast, the contribution of this paper is to generalize Dubois' definition and arithmetic operation of LR fuzzy numbers and then use this result to solve fuzzy matrix systems numerically. The importance of converting fuzzy linear system into two systems of linear equations is that any numerical approach suitable for system of linear equations may be implemented. In addition, since our model does not contain parameter $r, 0 \leq r \leq 1$, its numerical computation is relatively easy.

PRELIMINARIES

Definition 2.1: [2] A fuzzy number $\tilde{M}$ is said to be a LR fuzzy number if

$$
\mu_{\tilde{M}}(x) = \begin{cases} L\left(\dfrac{m-x}{\alpha}\right), & x \le m,\ \alpha > 0, \\[3mm] R\left(\dfrac{x-m}{\beta}\right), & x \ge m,\ \beta > 0, \end{cases}
$$

where m is the mean value of $\tilde{M}$, and α and β are left and right spreads, respectively. The function L(.), which is called left shape function satisfying: 1) L(x)=L(-x); 2) L(0)=1 and L(1)=0; 3) L(x) is a non increasing on $[0,+\infty)$.

The definition of a right shape function L(.) is usually similar to that of L(.). A LR fuzzy number $\tilde{M}$ is symbolically shown as $\tilde{M} = (m, \alpha, \beta)_{LR}$.

Noticing that $\alpha > 0, \beta > 0$ in Definition 2.1, which limits its applications, we extend the definition of LR fuzzy numbers as follows.

Definition 2.2: (Generalized LR fuzzy numbers) Let $\tilde{M} = (m, \alpha, \beta)_{LR}$, we define

1) if $\alpha < 0$, and $\beta > 0$, then $\tilde{M} = (m, 0, \text{Max}(-\alpha, \beta))_{LR}$, and

$$
\mu_{\tilde{M}}(x) = \begin{cases} 0, & x \le m, \\[3mm] R\left(\dfrac{x-m}{\max(-\alpha,\beta)}\right), & x \ge m. \end{cases}
$$

   a.

2) if $\alpha > 0$, and $\beta < 0$, then $\tilde{M} = \left(m, \text{Max}\left(-\alpha, \beta\right), 0\right)_{LR}$, and

$$\mu_{\tilde{M}}\left(x\right) = \begin{cases} L\left(\dfrac{m-x}{\max\left\{\alpha, -\beta\right\}}\right), & x \leq m, \\ 0, & x \geq m. \end{cases}$$

3) if $\alpha < 0$, and $\beta < 0$, then $\tilde{M} = \left(m, -\beta, -\alpha\right)_{LR}$, and

$$\mu_{\tilde{M}}\left(x\right) = \begin{cases} L\left(\dfrac{m-x}{-\beta}\right), & x \leq m, \\ R\left(\dfrac{x-m}{-\alpha}\right), & x \geq m. \end{cases}$$

For arbitrary LR fuzzy number $\tilde{M} = \left(m, \alpha, \beta\right)_{LR}$ and $\tilde{N} = \left(m, \gamma, \delta\right)_{LR}$, we have

1) $\tilde{M} + \tilde{N} = \left(m+n, \alpha+\gamma, \beta+\delta\right)_{LR}$.

2) $\lambda \tilde{N} = \begin{cases} \left(\lambda n, \lambda\gamma, \lambda\delta\right)_{LR}, & \lambda \geq 0, \\ \left(\lambda n, -\lambda\delta, -\lambda\gamma\right)_{RL}, & \lambda < 0. \end{cases}$

Definition 2.3: The matrix system

$$\begin{pmatrix} a_{11} & a_{12} & \cdots & a_{1n} \\ a_{21} & a_{22} & \cdots & a_{2n} \\ \cdots & \cdots & \cdots & \cdots \\ a_{m1} & a_{m2} & \cdots & a_{mn} \end{pmatrix} \begin{pmatrix} \tilde{x}_{11} & \tilde{x}_{12} & \cdots & \tilde{x}_{1p} \\ \tilde{x}_{21} & \tilde{x}_{22} & \cdots & \tilde{x}_{2p} \\ \cdots & \cdots & \cdots & \cdots \\ \tilde{x}_{n1} & \tilde{x}_{n2} & \cdots & \tilde{x}_{np} \end{pmatrix}$$

$$= \begin{pmatrix} \tilde{b}_{11} & \tilde{b}_{12} & \cdots & \tilde{b}_{1p} \\ \tilde{b}_{21} & \tilde{b}_{22} & \cdots & \tilde{b}_{2p} \\ \cdots & \cdots & \cdots & \cdots \\ \tilde{b}_{m1} & \tilde{b}_{m2} & \cdots & \tilde{b}_{mp} \end{pmatrix} \qquad (1)$$

where a_{ij} are crisp numbers and $\tilde{b}_{ij}$ are LR fuzzy numbers, is called a LR fuzzy matrix equation (LRFME).

Using matrix notation, we have

$$A\tilde{X} = \tilde{B} \tag{2}$$

A LR fuzzy numbers matrix

$$\tilde{X} = \left(\tilde{x}_{ij}\right)^T_{n \times p}, \quad \tilde{x}_{ij} = \left(x_{ij}, x^l_{ij}, x^r_{ij}\right)_{LR},$$

$$1 \le i \le n, \quad 1 \le j \le p$$

is called a solution of the LR fuzzy matrix systems if $\tilde{X}$ satisfies (2).

METHOD FOR SOLVING LRFME

In this section we investigate the LR fuzzy matrix system (2). Firstly, we propose a model for solving the LR fuzzy matrix system, i.e., convert it into two crisp systems of matrix equations. Then we define the LR fuzzy solution and give its solution representation to the original fuzzy matrix system. At last, the existence condition of the strong LR fuzzy solution to the original fuzzy matrix system is also discussed.

Extended Crisp Matrix Equations

By using arithmetic operations of LR fuzzy numbers, we extend the LR fuzzy matrix Equation (2) into two crisp matrix equations.

Theorem 3.1: The LR dual fuzzy linear Equation (2) can be extended into two crisp systems of linear equations as follows:

$$AX = B, \tag{3}$$

i.e.,

$$
\begin{pmatrix}
a_{11} & a_{12} & \cdots & a_{1n} \\
a_{21} & a_{22} & \cdots & a_{2n} \\
\cdots & \cdots & \cdots & \cdots \\
a_{m1} & a_{m2} & \cdots & a_{mn}
\end{pmatrix}
\begin{pmatrix}
x_{11} & x_{12} & \cdots & x_{1p} \\
x_{21} & x_{22} & \cdots & x_{2p} \\
\cdots & \cdots & \cdots & \cdots \\
x_{n1} & x_{n2} & \cdots & x_{np}
\end{pmatrix}
$$

$$
=
\begin{pmatrix}
b_{11} & b_{12} & \cdots & b_{1p} \\
b_{21} & b_{22} & \cdots & b_{2p} \\
\cdots & \cdots & \cdots & \cdots \\
b_{m1} & b_{m2} & \cdots & b_{mp}
\end{pmatrix}
$$

and

$$
S\begin{pmatrix} X^l \\ X^r \end{pmatrix} = \begin{pmatrix} B^l \\ B^r \end{pmatrix} = F , \tag{4}
$$

i.e.,

$$
\begin{pmatrix}
S_{11} & S_{12} & \cdots & S_{1.2n} \\
S_{21} & S_{22} & \cdots & S_{2.2n} \\
\cdots & \cdots & \cdots & \cdots \\
S_{2m.1} & S_{2m.2} & \cdots & S_{2m.2n}
\end{pmatrix}
\begin{pmatrix}
x_{11}^l & x_{12}^l & \cdots & x_{1p}^l \\
x_{21}^l & x_{22}^l & \cdots & x_{2p}^l \\
\cdots & \cdots & \cdots & \cdots \\
x_{n1}^l & x_{n2}^l & \cdots & x_{np}^l \\
x_{11}^r & x_{12}^r & \cdots & x_{1p}^r \\
x_{21}^r & x_{22}^r & \cdots & x_{2p}^r \\
\cdots & \cdots & \cdots & \cdots \\
x_{n1}^r & x_{n1}^r & \cdots & x_{np}^r
\end{pmatrix}
=
\begin{pmatrix}
b_{11}^l & b_{12}^l & \cdots & b_{1p}^l \\
b_{21}^l & b_{22}^l & \cdots & b_{2p}^l \\
\cdots & \cdots & \cdots & \cdots \\
b_{m1}^l & b_{m1}^l & \cdots & b_{mp}^l \\
b_{11}^r & b_{12}^r & \cdots & b_{1p}^r \\
b_{21}^r & b_{22}^r & \cdots & b_{2p}^r \\
\cdots & \cdots & \cdots & \cdots \\
b_{m1}^r & b_{m1}^r & \cdots & b_{mp}^r
\end{pmatrix} ,
$$

where $S_{ij}, 1 \leq i \leq 2m, 1 \leq j \leq 2n$ are determined as follows:

If $a_{ij} \geq 0$, then, $S_{ij} = a_{ij}$, $S_{m+i,\,n+j} = a_{ij}$ if $a_{ij} < 0$, then $S_{i,\,n+j} = a_{ij}$, $S_{m+i,\,n} = a_{ij}$, and any S_{kl} which is not determined by the above items is zero, $1 \leq k \leq 2m, 1 \leq l \leq 2n$.

Proof. Let $\tilde{X} = \left(\tilde{X}_1, \tilde{X}_2, ..., \tilde{X}_p \right)$, $\tilde{X}_j = \left(\left(x_{1j}, x_{1j}^l, x_{1j}^r \right)_{LR}, ..., \left(x_{nj}, x_{nj}^l, x_{nj}^r \right)_{LR} \right)^{\mathrm{T}}$

and $\tilde{B} = \left(\tilde{B}_1, \tilde{B}_2, ..., \tilde{B}_p \right)$, $\tilde{B}_j = \left(\left(b_{1j}, b_{1j}^l, b_{1j}^r \right)_{LR}, ..., \left(b_{nj}, b_{nj}^l, b_{nj}^r \right)_{LR} \right)^{\mathrm{T}}$.

Then the fuzzy matrix Equation (1) can be rewritten in the block forms $A\left(\tilde{X}_1, \tilde{X}_2, ..., \tilde{X}_p \right) = \left(\tilde{B}_1, \tilde{B}_2, ..., \tilde{B}_p \right)$. Thus the original system (1) is equivalent to the following fuzzy linear equations

$$A\tilde{X}_j = \tilde{B}_j, \quad 1 \leq j \leq p. \tag{5}$$

Now we consider the Equations (4). Let a_i be the ith row of matrix A, $1 \leq i \leq m$, we can represent $\left[A\tilde{X}_j \right]_i$ in the form $\left[A\tilde{X}_j \right]_i = a_i \tilde{X}_j, i = 1, 2, ..., m$.

Denoting $Q_i^+ = \left\{ a_{ik} : a_{ik} \geq 0 \right\}$ and $Q_i^- = \left\{ a_{ik} : a_{ik} < 0 \right\}$, we have

$$\left[A\tilde{X}_j \right]_i = \sum_{k \in Q_j^+} a_{ik} \tilde{x}_{kj} + \sum_{k \in Q_j^-} a_{ik} \tilde{x}_{kj}, \quad i = 1, 2, \cdots, m$$

i.e.,

$$\left[A\tilde{X}_j \right]_i = \left(\sum_{k \in Q_j^+} a_{ik} x_{kj} - \sum_{k \in Q_j^-} a_{ik} x_{kj}, \sum_{k \in Q_j^+} a_{ik} x_{kj}^l \right.$$
$$\left. - \sum_{k \in Q_j^-} a_{ik} x_{kj}^r, \sum_{k \in Q_j^+} a_{ik} x_{kj}^r - \sum_{k \in Q_j^-} a_{ik} x_{kj}^l \right)_{LR} \tag{6}$$

Consider the given LR fuzzy vector $\tilde{B}_j = \left(\left(b_{1j}, b^l_{1j}, b^r_{1j} \right)_{LR}, \ldots, \left(b_{nj}, b^l_{nj}, b^r_{nj} \right)_{LR} \right)^T$, we can write the system (2) as

$$\left(\begin{array}{c} \sum_{k \in Q^+_j} a_{ik} x_{kj} - \sum_{k \in Q^-_j} a_{ik} x_{kj}, \; \sum_{k \in Q^+_j} a_{ik} x^l_{kj} - \sum_{k \in Q^-_j} a_{ik} x^r_{kj}, \\[2mm] \sum_{k \in Q^+_j} a_{ik} x^r_{kj} - \sum_{k \in Q^-_j} a_{ik} x^l_{kj} \end{array} \right)_{LR} = \left(B_j, B^l_j, B^r_j \right)_{LR}$$

Suppose the system $A\tilde{X} = \tilde{B}_j, 1 \leq j \leq p$, has a solution. Then, the corresponding mean value $X_j = \left(x_{1j}, x_{2j}, \ldots, x_{nj} \right)^T$ of the solution must lie in the following linear system

$$\begin{pmatrix} a_{11} & a_{12} & \cdots & a_{1n} \\ a_{21} & a_{22} & \cdots & a_{2n} \\ \cdots & \cdots & \cdots & \cdots \\ a_{m1} & a_{m2} & \cdots & a_{mn} \end{pmatrix} \begin{pmatrix} x_{1j} \\ x_{2j} \\ \cdots \\ x_{nj} \end{pmatrix} = \begin{pmatrix} b_{1j} \\ b_{2j} \\ \cdots \\ b_{nj} \end{pmatrix}. \tag{7}$$

Meanwhile, the left spread $X^l_j = \left(x^l_{1j}, x^l_{2j}, \ldots, x^l_{nj} \right)^T$ and the right spread $X^r_j = \left(x^r_{1j}, x^r_{2j}, \ldots, x^r_{nj} \right)^T$ of the solution can be derived from solving the following crisp linear system

$$\begin{pmatrix} S_{11} & S_{12} & \cdots & S_{1,2n} \\ S_{21} & S_{22} & \cdots & S_{2,2n} \\ \cdots & \cdots & \cdots & \cdots \\ S_{2m,1} & S_{2m,2} & \cdots & S_{2m,2n} \end{pmatrix} \begin{pmatrix} x^l_{1j} \\ \vdots \\ x^l_{nj} \\ x^r_{1j} \\ \vdots \\ x^r_{nj} \end{pmatrix} = \begin{pmatrix} b^l_{1j} \\ \vdots \\ b^l_{nj} \\ b^r_{1j} \\ \vdots \\ b^r_{nj} \end{pmatrix}. \tag{8}$$

Finally, we restore the Equation (5) and obtain above matrix Equations (3) and (4).

The proof is completed.

Computing Model Matrix Equations

In order to solve the original fuzzy linear Equation (2), we need to consider crisp matrix Equations (3) and (4). Since Equations (3) and (4) are crisp, their computation is relatively easy.

In general [14], the minimal solutions of matrix systems (3) and (4) can be expressed uniformly by

$$X = A^+B \tag{9}$$

and

$$\begin{pmatrix} X^l \\ X^r \end{pmatrix} = S^+ \begin{pmatrix} B^l \\ B^r \end{pmatrix} = S^+F \tag{10}$$

respectively, no matter the Equations (3) and (4) are consistent or not.

It seems that we have obtained the solution of the original fuzzy linear system (2) as follows:

$$\tilde{X} = \left(X, X^l, X^r \right)_{LR}$$
$$= \left(A^+d, \left(I_n \ O \right) S^+F, \left(O \ I_n \right) S^+F \right)_{LR} \tag{11}$$

But the solution vector may still not be an appropriate LR fuzzy numbers vector except for $S^+F \geq 0$. So we give the definition of the minimal LR fuzzy solution to the Equation (2) as follows:

Definition 3.1: Let $\tilde{X} = \left(x_{ij}, x_{ij}^l, x_{ij}^r \right)_{LR}$ $1 \leq i \leq n, 1 \leq j \leq p$. If $X = \left(x_{ij} \right)_{n \times p}$ is the minimal solution of Equation (3), $X^l = \left(x_{ij} \right)_{n \times p}$ and $X^r = \left(x_{ij}^r \right)_{n \times p}$ are minimal solution of Equation (4) such that $X^l \geq 0, X^r \geq 0$ then we call $\tilde{X} = \left(x, x^l, x^r \right)_{LR}$ is a strong LR fuzzy solution of Equation (2). Otherwise, it is a weak LR fuzzy solution of Equation (2) given by

$$
\tilde{x}_{ij} = \begin{cases}
\left(x_{ij}, x_{ij}^l, x_{ij}^r \right)_{LR}, & x_{ij}^l > 0, \, x_{ij}^r > 0, \\
\left(x_{ij}, 0, \max\left\{ -x_{ij}^l, x_{ij}^r \right\} \right)_{LR}, & x_{ij}^l < 0, \, x_{ij}^r > 0, \\
\left(x_{ij}, \max\left\{ x_{ij}^l, -x_{ij}^r \right\}, 0 \right)_{LR}, & x_{ij}^l > 0, \, x_{ij}^r < 0, \\
\left(x_{ij}, -x_{ij}^r, -x_{ij}^l \right)_{LR}, & x_{ij}^l < 0, \, x_{ij}^r < 0.
\end{cases}
$$

$$1 \leq i \leq n, \ 1 \leq j \leq p. \tag{12}$$

A Sufficient Condition of Strong Fuzzy Solution

The key points to make the solution vector being a LR fuzzy solution are $X^l \geq 0$, and $X^r \geq 0$. Since

$$X^l = \left(I_n, O \right) S^+ F, \, X^r = \left(O I_n \right) S^+ F$$, we know that the non-negativities of X

and X^r are equivalent to the condition $S^+ \geq 0$ now that $\begin{pmatrix} X^l \\ X^r \end{pmatrix} \geq 0$ is known.

By the above analysis, we have the following result.

Theorem 3.2: Let A belong to $R^{m \times n}$. If S^+ is nonnegative, the solution of the LR fuzzy matrix system (2) is expressed by

$$\tilde{X} = \left(X, X^l, X^r \right)_{LR}$$
$$= \left(A^+ d, \left(I_n\ O \right) S^+ F, \left(O\ I_n \right) S^+ F \right)_{LR} \qquad (13)$$

and it admits a strong minimal LR fuzzy solution.

The following Theorem gives a result for such S^+ to be nonnegative.

Theorem 3.3: [15] Let S be an $2p \times 2p$ nonnegative matrix with rank r. Then the following assertions are equivalent:

1) $S^+ \geq 0$;

2) There exists a permutation matrix P, such that PS has the form

$$PS = \begin{pmatrix} Q_1 \\ \vdots \\ Q_r \\ O \end{pmatrix}$$ where each Q_i has rank 1 and the rows of Q_i are or-

thogonal to the rows of Q_i, whenever $i \neq j$, the zero matrix may be absent.

3) $S^+ = \begin{pmatrix} HE^T & HF^T \\ HF^T & HE^T \end{pmatrix}$ for some positive diagonal matrix H. In this case,

$$(E+F)^+ = H(E+F)^T, (E-F)^+ = H(E-F)^T$$

NUMERICAL EXAMPLES

In this section, we work out two numerical examples to illustrate the proposed method.

Example 4.1: Consider the fuzzy matrix systems:

$$\begin{pmatrix} 1 & 0 & -1 \\ 1 & -1 & 0 \\ 2 & 1 & 1 \end{pmatrix} \begin{pmatrix} \tilde{x}_{11} & \tilde{x}_{12} \\ \tilde{x}_{21} & \tilde{x}_{22} \\ \tilde{x}_{31} & \tilde{x}_{32} \end{pmatrix} = \begin{pmatrix} (2,1,1)_{LR} & (3,2,1)_{LR} \\ (2,1,2)_{LR} & (2,1,2)_{LR} \\ (6,3,2)_{LR} & (5,2,3)_{LR} \end{pmatrix}$$

The coefficient matrix A is nonsingular and the extended matrix S is singular. By the Theorem 3.1., the mean value x, the left spread x^l and the right spread x^r of solution are obtained from

$$\begin{pmatrix} 1 & 0 & -1 \\ 1 & -1 & 0 \\ 2 & 1 & 1 \end{pmatrix} \begin{pmatrix} x_{11} & x_{12} \\ x_{21} & x_{22} \\ x_{31} & x_{32} \end{pmatrix} = \begin{pmatrix} 2 & 3 \\ 2 & 2 \\ 6 & 5 \end{pmatrix}$$

and

$$\begin{pmatrix} 1 & 1 & 0 & 0 & 0 & 1 \\ 1 & 0 & 0 & 0 & 1 & 0 \\ 2 & 1 & 1 & 0 & 0 & 0 \\ 0 & 0 & 1 & 1 & 1 & 0 \\ 0 & 1 & 0 & 1 & 0 & 0 \\ 0 & 0 & 0 & 2 & 1 & 1 \end{pmatrix} \begin{pmatrix} x_{11}^l & x_{12}^l \\ x_{21}^l & x_{22}^l \\ x_{31}^l & x_{32}^l \\ x_{11}^r & x_{12}^r \\ x_{21}^r & x_{22}^r \\ x_{31}^r & x_{32}^r \end{pmatrix} = \begin{pmatrix} 1 & 2 \\ 1 & 1 \\ 3 & 2 \\ 1 & 1 \\ 2 & 2 \\ 2 & 3 \end{pmatrix}$$

Thus, we have

$$x = A^+ B = \begin{pmatrix} 2.5000, & 2.5000 \\ 0.5000, & 0.5000 \\ 0.5000, & -0.500 \end{pmatrix}$$

and

$$\begin{pmatrix} x^l \\ x^r \end{pmatrix} = S^+ \begin{pmatrix} B^l \\ B^r \end{pmatrix} = \begin{pmatrix} 0.8333, & 0.9639 \\ 1.1667, & 0.8194 \\ 0.1667, & -0.1806 \\ 0.8333, & 1.0139 \\ 0.1667, & 0.0694 \\ 0.1667, & 1.0694 \end{pmatrix}.$$

Since x_{23}^l is negative, according to Definition 3.1., the LR fuzzy approximate solution of the original fuzzy matrix system is

$$\tilde{X} = \begin{pmatrix} \tilde{x}_{11} & \tilde{x}_{12} \\ \tilde{x}_{21} & \tilde{x}_{22} \\ \tilde{x}_{31} & \tilde{x}_{32} \end{pmatrix} = \begin{pmatrix} (2.5000, & 0.8333, & 0.8333)_{LR} & (2.5000, & 0.7639, & 1.0139)_{LR} \\ (0.5000, & 1.1667, & 0.1667)_{LR} & (0.5000, & 0.8194, & 0.0694)_{LR} \\ (0.5000, & 0.1667, & 0.1667)_{LR} & (-0.500, & 0.0000, & 1.0694)_{LR} \end{pmatrix}$$

and it admits a strong LR fuzzy solution.

Example 4.2: Consider the following matrix systems:

$$\begin{pmatrix} 1 & 1 & 1 \\ 1 & 1 & -1 \end{pmatrix} \begin{pmatrix} \tilde{x}_{11} & \tilde{x}_{12} \\ \tilde{x}_{21} & \tilde{x}_{22} \\ \tilde{x}_{31} & \tilde{x}_{32} \end{pmatrix} = \begin{pmatrix} (10, & 3, & 5)_{LR} & (6, & 2, & 2)_{LR} \\ (4, & 2, & 1)_{LR} & (5, & 3, & 1)_{LR} \end{pmatrix}.$$

By the Theorem 3.1., the mean value x of solution lies in the following crisp matrix system

$$\begin{pmatrix} 1 & 1 & 1 \\ 1 & 1 & -1 \end{pmatrix} \begin{pmatrix} x_{11} & x_{12} \\ x_{21} & x_{22} \\ x_{31} & x_{32} \end{pmatrix} = \begin{pmatrix} 10 & 6 \\ 4 & 5 \end{pmatrix}.$$

Meanwhile, the left spread x^l and the right spread x^r of solution are obtained by solving the following crisp matrix system

$$\begin{pmatrix} 1 & 1 & 1 & 0 & 0 & 0 \\ 1 & 1 & 0 & 0 & 0 & 1 \\ 0 & 0 & 0 & 1 & 1 & 1 \\ 0 & 0 & 1 & 1 & 1 & 0 \end{pmatrix} \begin{pmatrix} x_{11}^l & x_{12}^l \\ x_{21}^l & x_{22}^l \\ x_{31}^l & x_{32}^l \\ x_{11}^r & x_{12}^r \\ x_{21}^r & x_{22}^r \\ x_{31}^r & x_{32}^r \end{pmatrix} = \begin{pmatrix} 3 & 2 \\ 2 & 3 \\ 5 & 2 \\ 1 & 1 \end{pmatrix} .$$

Thus, we have

$$x = A^+ B = \begin{pmatrix} 3.5000, & 2.7500 \\ 3.5000, & 2.7500 \\ 3.0000, & 0.5000 \end{pmatrix}$$

and

$$\begin{pmatrix} x^l \\ x^r \end{pmatrix} = S^+ \begin{pmatrix} B^l \\ B^r \end{pmatrix} = \begin{pmatrix} 0.7917, & 0.9167 \\ 0.7917, & 0.9167 \\ 0.1667, & 0.1667 \\ 1.0417, & 0.4167 \\ 1.0417, & 0.4167 \\ 1.1667, & 1.1667 \end{pmatrix} > 0.$$

By Definition 3.1., the original fuzzy system has a strong LR fuzzy approximate solution given by

$$\tilde{X} = \begin{pmatrix} \tilde{x}_{11} & \tilde{x}_{12} \\ \tilde{x}_{21} & \tilde{x}_{22} \\ \tilde{x}_{31} & \tilde{x}_{32} \end{pmatrix} = \begin{pmatrix} (3.50, & 0.792, & 1.042)_{LR} & (2.75, & 0.917, & 0.417)_{LR} \\ (3.50, & 0.792, & 1.042)_{LR} & (2.75, & 0.917, & 0.417)_{LR} \\ (3.00, & 0.167, & 1.667)_{LR} & (0.50, & 0.167, & 1.167)_{LR} \end{pmatrix} .$$

CONCLUSIONS

In this work we proposed a general model for solving the fuzzy matrix equation $A\tilde{X} = \tilde{B}$ where A is an m × n crisp matrix and $\tilde{B}$ is a n × p arbitrary LR fuzzy numbers matrix. We converted the fuzzy matrix system into two crisp matrix equations and obtained the LR fuzzy solution to the original fuzzy system by solving crisp matrix equations. Moreover, the existence condition of strong LR fuzzy solution was studied. Numerical examples showed that our method is effective to solve LR fuzzy matrix equations.

ACKNOWLEDGMENTS

The authors are very thankful for the reviewer's helpful suggestions to improve the paper. This research was financially supported by the National Natural Science Foundation of China (Grant No. 71061013) and the Youth Scientific Research Ability Promotion Project of Northwest Normal University (NWNU-LKQN-11-20).

REFERENCES

1. L A. Zadeh, "Fuzzy Sets," Information and Control, Vol. 8, No. 3, 1965, pp. 338-353.doi:10.1016/S0019-9958(65)90241-X
2. D. Dubois, and H. Prade, "Operations on Fuzzy Numbers," International Journal of Systems Science, Vol. 9, No. 3, 1978, pp. 613-626. doi:10.1080/00207727808941724
3. M. Friedman, M. Ma and A. Kandel, "Fuzzy Linear Systems," Fuzzy Sets and Systems, Vol. 96, No. 2, 1998, pp. 201-209. doi:10.1016/S0165-0114(96)00270-9
4. M. Friedman, M. Ma and A. Kandel, "Duality in Fuzzy Linear Systems," Fuzzy Sets and Systems, Vol. 109, No. 1, 2000, pp. 55-58. doi:10.1016/S0165-0114(98)00102-X
5. T. Allahviranloo, "Numerical Methods for Fuzzy System of Linear Equations," Applied Mathematics and Computation, Vol. 155, No. 2, 2004, pp. 493-502. doi:10.1016/S0096-3003(03)00793-8

6. B. Asady, S. Abbasbandy and M. Alavi, "Fuzzy General Linear Systems," Applied Mathematics and Computation, Vol. 169, No. 1, 2005, pp. 34-40.doi:10.1016/j.amc.2004.10.042

7. K. Wang and B. Zheng, "Inconsistent Fuzzy Linear Systems," Applied Mathematics and Computation, Vol. 181, No. 2, 2006, pp. 973-981. doi:10.1016/j.amc.2006.02.019

8. M. Dehghan, B. Hashemi and M. Ghatee, "Solution of the Full Fuzzy Linear Systems Using Iterative Techniques," Chaos, Solitons & Fractals, Vol. 34, No. 2, 2007, pp. 316- 336. doi:10.1016/j.chaos.2006.03.085

9. S. Abbasbandy, M. Otadi and M. Mosleh, "Minimal Solution of General Dual Fuzzy Linear Systems," Chaos, Solitons & Fractals, Vol. 37, No. 4, 2008, pp. 638-652.doi:10.1016/j.chaos.2006.10.045

10. X. B. Guo and Z. T. Gong, "Undetermined Coefficient Method for Solving Semi-Fuzzy Matrix Equation," Proceedings of 9th International Conference on Machine Learning and Cybernetics, Qingdao, 11-14 July 2010, pp. 596-600.

11. Z. T. Gong and X. B. Guo, "Inconsistent Fuzzy Matrix Equations and Its Fuzzy Least Squares Solutions," Applied Mathematical Modelling, Vol. 35, No. 1, 2011, pp. 1456-1469. doi:10.1016/j.apm.2010.09.022

12. X. B. Guo, "Approximate Solution of Fuzzy Sylvester Matrix Equations," The 7th International Conference on Computational Intelligence and Security, Sanya, 3-4 December 2011, pp. 52-57.

13. X. B. Guo and D. Q. Shang, "Fuzzy Symmetric Solutions of Fuzzy Matrix Equations," Advances in Fuzzy Systems, 2012, Article ID: 318069. doi:10.1155/2012/318069

14. X. D. Zhang, "Matrix Analysis and Its Applications," Tsinghua University and Springer Press, Beijing, 2004.

15. A. Berman and R. J. Plemmons, "Nonnegative Matrices in the Mathematical Sciences," Academic Press, New York, 1979.

CITATION

X. Guo and D. Shang, "Approximate Solution of Fuzzy Matrix Equations with LR Fuzzy Numbers," Advances in Pure Mathematics, Vol. 2 No. 6, 2012, pp. 373-378. doi: 10.4236/apm.2012.26056.

Generalizations of a Matrix Inequality

Lingzhi Zhao[1], Jun Yuan[2], and Yunfeng Cai[3]
[1]School of Mathematics and Information Technology, Nanjing Xiaozhuang University, Nanjing
[2]College of Teacher Education, Nanjing Xiaozhuang University, Nanjing
[3]College of Science, Nanjing University of Posts and Telecommunications, Nanjing

INTRODUCTION

The well-known Brunn-Minkowski inequality is one of the most important inequalities in geometry. There are many other interesting results
related to the Brunn-Minkowski inequality (see [1-8]). The matrix form
of the Brunn-Minkowski inequality (see [9, 10]) asserts that if A and B
are two positive definite matrices of order n and $o < \lambda < 1$, then

$$\left|\lambda A + (1-\lambda) B\right|^{\frac{1}{n}} \geq \lambda |A|^{\frac{1}{n}} + (1-\lambda)|B|^{\frac{1}{n}},$$

$$(1)$$

with equality if and only if $A = cB (c \geq 0)$, where $|A|$ denotes the deter-
minant of A.

Let $R^{n\times n}$ denote the set of $n\times n$ real symmetry matrices. Let I_n denote
$n\times n$ unit matrix. We use the notation $A > 0 (A \geq 0)$ if A is a positive
definite (positive semi-definite) matrix, and A^* denotes the transpose
of A. Let $A, B \in R^{n\times n}$, then $A > B (A \geq B)$ if and only if $A - B > 0 (A - B \geq 0)$
If $A \in R^{n\times n}$, then there exists a unitary matrix U such as

$$A = U^* [\lambda_1, \cdots, \lambda_n] U,$$

where $[\lambda_1, L, \lambda_n]$ is a diagonal matrix $(\lambda_i \delta_{ij})$, and λ_1, L, λ_n are the eigenvalues of A, each appearing as its multiplicity. Assume now that $f(\lambda_i) \in R$ is well defined. Then $f(A)$ may be defined by (see e.g. [11, p. 71] or [12, p. 90])

$$f(A) = U^* \big[f(\lambda_1), \cdots, f(\lambda_n) \big] U.$$

(2)

In this paper, some new generalizations of the matrix form of the Brunn-Minkowski inequality are presented. One of our main results is the following theorem.

Theorem 1.1: Let A, B be positive definite commuting matrix of order n with eigenvalues in the interval I. If f is a positive concave function on I and $0 < \lambda < 1$, then

$$\left| f(\lambda A + (1-\lambda)B) \right|^{\frac{1}{n}} \geq \lambda \left| f(A) \right|^{\frac{1}{n}} + (1-\lambda) \left| f(B) \right|^{\frac{1}{n}}$$

(3)

with equality if and only if f is linear and $f(A) = cf(B)(c \geq 0)$.
Let $A, B \in R^{n \times n}$, if $A \geq B$. We can define the determinant differences function of A and by B.by.

$$D_d(A, B) = |A| - |B|.$$

The following theorem gives another generalization of (1).

Theorem 1.2: Let A, B be positive definite commuting matrix of order n with eigenvalues in the interval I and $1 \in I$. Let f be a positive function on I and a and b be two nonnegative real numbers such that

$$f(A) > af(I_n), \quad f(B) > bf(I_n).$$

Then

$$D_d\big(f(A) + f(B), (a+b)f(I_n)\big)^{\frac{1}{n}} \geq D_d\big(f(A), af(I_n)\big)^{\frac{1}{n}} + D_d\big(f(B), bf(I_n)\big)^{\frac{1}{n}}$$

(4)

with equality if and only if $\alpha^{-1}f(A) = b^{-1}f(B)$

Remark 1. Let f(t)=t in Theorem 1.1 or let V f(t)=t and a=b=0 in Theorem 1.2. We can both obtain (1). Hence Theorem 1.1 and Theorem 1.2 are generalizations of (1).

PROOFS OF THEOREMS

To prove the theorems, we need the following lemmas:

Lemma 2.1: ([13], p.472) Let $A, B \in R^{n \times n}$, $A > B > 0$. Then

$$|A| > |B|.$$

Lemma 2.2: ([13], p.50) Let $A, B \in R^{n \times n}$ $A > 0, B > 0$, If A and B are commute, then exists a unitary matrix U such that

$$U^*AU = [a_1, a_2, \cdots, a_n] \text{ and } U^*BU = [b_1, b_2, \cdots, b_n].$$

Lemma 2.3: ([14], p.35) Let $x_i \geq 0, y_i \geq 0 (i = 1, 2, L, n)$. Then

$$\left(\prod_{i=1}^{n} x_i\right)^{\frac{1}{n}} + \left(\prod_{i=1}^{n} y_i\right)^{\frac{1}{n}} \leq \left(\prod_{i=1}^{n} (x_i + y_i)\right)^{\frac{1}{n}},$$

with equality if and only if $x_i = v y_i$, where V is a constant.

This is a special case of Maclaurin's inequality.

Proof of Theorem 1.1

Since A and B are commuted, by lemma 2.2, there exists a unitary matrix U such that

$$A = U^*[a_1, a_2, \cdots, a_n]U \text{ and } B = U^*[b_1, b_2, \cdots, b_n]U.$$

Hence,

$$\lambda A + (1-\lambda)B = U^* \left[\lambda a_1 + (1-\lambda)b_1, \lambda a_2 + (1-\lambda)b_2, \cdots, \lambda a_n + (1-\lambda)b_n \right] U.$$

By (2), we have

$$f(A) = U^* \left[f(a_1), f(a_2), \cdots, f(a_n) \right] U,$$

$$f(B) = U^* \left[f(b_1), f(b_2), \cdots, f(b_n) \right] U,$$

and

$$f(\lambda A + (1-\lambda)B) = U^* \left[f(\lambda a_1 + (1-\lambda)b_1), f(\lambda a_2 + (1-\lambda)b_2), f(\lambda a_n + (1-\lambda)b_n) \right] U.$$

Since f is a concave function, by lemma 2.3, we get

$$\left| f(\lambda A + (1-\lambda)B) \right|^{\frac{1}{n}} = \left(\prod_{i=1}^{n} f(\lambda a_i + (1-\lambda)b_i) \right)^{\frac{1}{n}}$$

$$\geq \left(\prod_{i=1}^{n} \left[\lambda f(a_i) + (1-\lambda)f(b_i) \right] \right)^{\frac{1}{n}} \tag{5}$$

$$\geq \lambda \left(\prod_{i=1}^{n} f(a_i) \right)^{\frac{1}{n}} + (1-\lambda)\left(\prod_{i=1}^{n} f(b_i) \right)^{\frac{1}{n}} \tag{6}$$

$$= \lambda \left| f(A) \right|^{\frac{1}{n}} + (1-\lambda)\left| f(B) \right|^{\frac{1}{n}}.$$

Now we consider the conditions of equality holds. Since f is a concave function, the equality of (5) holds if and only if f is linear. By the equality of Lemma 2.3, the equality of (6) holds if and only if f (a$_i$) =c f (b$_i$) which means f(A)=c f(B). So the equality of (3) holds if and only if f is linear and f (A) =c f (B) $(c \geq 0)$. This completes the proof of the Theorem 1.1.

Applying the arithmetic-geometric mean inequality to the right side of (3), we get the following corollary.

Corollary 2.4: Let A, B be positive definite commuting matrix of order n with eigenvalues in the interval I. If f is a positive concave function on I and $0 < \lambda < 1$, then

$$\left| f\left(\lambda A + (1-\lambda)B\right) \right| \geq \left| f(A) \right|^{\lambda} \left| f(B) \right|^{1-\lambda},$$

with equality if and only if A=B.

Taking for f(t)=t in Corollary 2.4, we obtain the Fan Ky concave theorem.

Proof of Theorem 1.2

As in the proof of Theorem 1.1, since A and B are commuted, by lemma 2.2, there exists a unitary matrix U such that

$$f(A) = U^{*}\left[f(a_{1}), f(a_{2}), \cdots, f(a_{n}) \right] U$$

and

$$f(B) = U^{*}\left[f(b_{1}), f(b_{2}), \cdots, f(b_{n}) \right] U.$$

So

$$\left| f(A) \right| = \prod_{i=1}^{n} f(a_{i}), \quad \left| f(B) \right| = \prod_{i=1}^{n} f(b_{i}),$$

$$\left| f(A) + f(B) \right| = \prod_{i=1}^{n}\left(f(a_{i}) + f(b_{i}) \right).$$

It is easy to see that (4) holds if and only if

$$\left(\prod_{i=1}^{n} \left(f(a_i) + f(b_i) \right) - \left[(a+b) f(1) \right]^n \right)^{\frac{1}{n}}$$

$$\geq \left(\prod_{i=1}^{n} f(a_i) - \left[af(1) \right]^n \right)^{\frac{1}{n}} + \left(\prod_{i=1}^{n} f(b_i) - \left[bf(1) \right]^n \right)^{\frac{1}{n}}$$

(7)

Since $f(A) > af(I_n), f(B) > bf(I_n)$, by Lemma 2.1, we have

$$\prod_{i=1}^{n} f(a_i) > \left[af(1) \right]^n, \quad \prod_{i=1}^{n} f(b_i) > \left[bf(1) \right]^n.$$

Now we prove (7). Put

$$X^n = \prod_{i=1}^{n} f(a_i) - \left[af(1) \right]^n, \quad Y^n = \prod_{i=1}^{n} f(b_i) - \left[bf(1) \right]^n.$$

Then

$$X^n + \left[af(1) \right]^n = \prod_{i=1}^{n} f(a_i), \quad Y^n + \left[bf(1) \right]^n = \prod_{i=1}^{n} f(b_i).$$

Applying Minkowski inequality, we have

$$\left((X+Y)^n + \left[(a+b) f(1) \right]^n \right)^{\frac{1}{n}} \leq \left(X^n + \left[af(1) \right]^n \right)^{\frac{1}{n}} + \left(Y^n + \left[bf(1) \right]^n \right)^{\frac{1}{n}}$$

$$= \left(\prod_{i=1}^{n} f(a_i) \right)^{\frac{1}{n}} + \left(\prod_{i=1}^{n} f(b_i) \right)^{\frac{1}{n}}$$

Using the Lemma 2.3 to the right of the above inequlity, we obtain

$$\left((X+Y)^n + \left[(a+b) f(1) \right]^n \right)^{\frac{1}{n}} \leq \left(\prod_{i=1}^{n} \left(f(a_i) + f(b_i) \right) \right)^{\frac{1}{n}},$$

which implies that

$$(X+Y)^n \leq \prod_{i=1}^{n}(f(a_i)+f(b_i))-[(a+b)f(1)]^n.$$

It follows that

$$X+Y \leq \left(\prod_{i=1}^{n}(f(a_i)+f(b_i))-[(a+b)f(1)]^n\right)^{\frac{1}{n}},$$

which is just the inequality (7).

By the equality conditions of Minkowski inequality and Lemma 2.3, the equality (1.4) holds if and only if $a^{-1}f(\alpha_i) = b^{-1}f(b_i)$, which means $a^{-1}f(A) = b^{-1}f(B)$. Thus we complete the proof of Theorem 1.2.

Taking for f(t)=t in Theorem 1.2, we obtain the following corollary.

Corollary 2.5: [7] Let A, B be positive definite commuting matrix of order n and a and b be two nonnegative real numbers such that

$$A > aI_n, \quad B > bI_n.$$

Then

$$\left(|A+B|-|(a+b)I_n|\right)^{\frac{1}{n}} \geq \left(|A|-|aI_n|\right)^{\frac{1}{n}} + \left(|B|-|bI_n|\right)^{\frac{1}{n}}$$

with equality if and only if $a^{-1}A = b^{-1}B$.

ACKNOWLEDGMENTS

The authors are most grateful to the referee for his valuable suggestions. And the authors would like to acknowledge the support from the National Natural Science Foundation of China (11101216,11161024), Qing Lan Project and the Nanjing Xiaozhuang University (2010KYQN24, 2010KYYB13).

REFERENCES

1. J. Bakelman, "Convex Analysis and Nonlinear Geometric Elliptic Equations," Springer, Berlin, 1994. http://dx.doi.org/10.1007/978-3-642-69881-1
2. C. Borell, "The Brunn-Minkowski Inequality in Gauss Space," Inventiones Mathematicae, Vol. 30, No. 2, 1975, pp. 202-216. http://dx.doi.org/10.1007/BF01425510
3. C. Borell, "Capacitary Inequality of the Brunn-Minkowski Inequality Type," Mathematische Annalen, Vol. 263, No. 2, 1993, pp. 179-184. http://dx.doi.org/10.1007/BF01456879
4. K. Fan, "Some Inequality Concerning Positive-Denite Hermitian Matrices," Mathematical Proceedings of the Cambridge Philosophical Society, Vol. 51, No. 3, 1958, pp. 414-421.http://dx.doi.org/10.1017/S0305004100030413
5. R. J. Gardner and P. Gronchi, "A Brunn-Minkowski Inequality for the Integer Lattice," Transactions of the American Mathematical Society, Vol. 353, No. 10, 2001, pp. 3995-4024. http://dx.doi.org/10.1090/S0002-9947-01-02763-5
6. R. J. Gardner, "The Brunn-Minkowski Inequality," Bulletin of the American Mathematical Society, Vol. 39, No. 3, 2002, pp. 355-405. http://dx.doi.org/10.1090/S0273-0979-02-00941-2
7. G. S. Leng, "The Brunn-Minkowski Inequality for Volume Differences," Advances in Applied Mathematics, Vol. 32, No. 3, 2004, pp. 615-624.http://dx.doi.org/10.1016/S0196-8858(03)00095-2
8. R. Osserman, "The Brunn-Minkowski Inequality for Multiplictities," Inventiones Mathematicae, Vol. 125, No. 3, 1996, pp. 405-411.http://dx.doi.org/10.1007/s002220050081
9. E. V. Haynesworth, "Note on Bounds for Certain Determinants," Duke Mathematical Journal, Vol. 24, No. 3, 1957, pp. 313- 320. http://dx.doi.org/10.1215/S0012-7094-57-02437-7
10. E. V. Haynesworth, "Bounds for Determinants with Positive Diagonals," Transactions of the American Mathematical Society, Vol. 96, No. 3, 1960, pp. 395-413. http://dx.doi.org/10.1090/S0002-9947-1960-0120242-1
11. M. Marcus and H. Minc, "A Survey of Matrix Theory and Inequalities," Allyn and Bacon, Boston, 1964.
12. R. Bellman, "Introduction to Matrix Analysis," McGraw-Hill, New York, 1960.
13. R. Horn and C. R. Johnson, "Matrix Analysis," Cambridge University Press, New York, 1985. http://dx.doi.org/10.1017/CBO9780511810817
14. E. F. Beckenbach and R. Bellman, "Inequalities," Springer, Berlin, 1961.

CITATION

L. Zhao, J. Yuan and Y. Cai, "Generalizations of a Matrix Inequality," Applied Mathematics, Vol. 5 No. 3, 2014, pp. 337-341. doi: 10.4236/am.2014.53034.

On Two Problems for Matrix Polytopes

Şerife Yılmaz and Taner Büyükköro lu
Department of Mathematics, Faculty of Science,
Anadolu University, Eskisehir, Turkey

5

ABSTRACT

We consider two problems from stability theory of matrix polytopes: the existence of common quadratic Lyapunov functions and the existence of a stable member. We show the applicability of the gradient algorithm and give a new sufficient condition for the second problem. A number of examples are considered.

INTRODUCTION

Consider the switched system

$$\dot{x}(t) = Ax(t), \quad A \in \{A_1, A_2, \cdots, A_N\} \tag{1}$$

where $\dot{x}(t) \in \mathbb{R}^n, t \geq 0$. In Equation (1), the matrix A switches among N matrices $A_1, A_2, \dots A_N$.

Switching signal $\sigma(t)$ is piecewise continuous from the right function $\sigma : [0,\infty) \to \{1,2\dots,N\}$ and the switching times are arbitrary. For the switched system (1) with initial condition $x(0)=x_0$ and with switching signal $\sigma(t)$ denotes the solution by $x(t,x_0,\sigma(.))$.

Definition 1: The origin is uniformly asymptotically stable (UAS) for the system (1) if for every $\varepsilon > 0$ there exists $\delta > 0$ such that for every signal $\sigma(t)$ and initial state x_0 with $\|x_0\| < \delta,$, the inequality $\|x(t, x_0, \sigma(.))\| < \varepsilon$ is satisfied for all $t > 0$ and uniformly on Σ (.)

$$\lim_{t \to \infty} x\left(t, x_0, \sigma(\cdot)\right) = 0.$$

If all systems in (1) share a common quadratic Lyapunov function (CQLF) $V(x) = x^T P x$ then the switched system is UAS (T denotes the transpose).

In this case there exists a common P>0 such that

$$A_i^T P + P A_i < 0 \quad (i = 1, 2, \cdots, N)$$

(2)

and P is called a common solution to the set of Lyapunov matrix inequalities (2).

The problem of existence of common positive definite solution P of (2) has been studied in a lot of works (see [1] - [9] and references therein). Numerical solution for common P via nondifferentiable convex optimization has been discussed in [10].

In the first part of the paper, the problem of existence of CQLF is investigated by Kelley's method. This method is applied when CQLF problem is treated as a convex optimization problem.

Second part of the paper is devoted to the following question:

Let $B \subset \mathbb{R}^l$ be a compact, for $q \in B$ the matrix A(q) is a real $n \times n$ matrix. Is there a Hurwitz stable member (all eigenvalues lie in the open left half plane) in the family

$$\left\{ A(q) : q \in B \right\}$$

or equivalently is there $q^* \in B$ such that $A\left(q^*\right)$ is stable? This problem is one of the hard and important problems in control theory (see [11]). Numerical solution of this problem is considered in [12] . In this paper we reduce this problem to a non-convex optimization problem.

COMMON QUADRATIC LYAPUNOV FUNCTION

For the switched system

$$\dot{x} = \{A_1, A_2, \cdots, A_N\}x$$

consider the problem of determination of CQLF $V(x) = x^{\mathrm{T}}Px$ where $P > 0$. We are going to investigate it by Kelley's cutting-plane method. This method gives new sufficient condition (Theorem 2) and new algorithm (Algorithm 1) which is more effective in comparison with the algorithm from [10] .

Consider the problem of existence of a common $P > 0$ such that

$$A_i^{\mathrm{T}} P + P A_i < 0 \quad (i = 1, 2, \cdots, N). \tag{3}$$

Let $x \in \mathbb{R}^r$ be $x = (x_1, x_2, \ldots, x_r)$ and P be an $n \times n$ symmetric matrix defined as

$$P = P(x) = \begin{pmatrix} x_1 & x_2 & \cdots & x_n \\ x_2 & x_{n+1} & \cdots & x_{2n-1} \\ \vdots & \vdots & \ddots & \vdots \\ x_n & x_{2n-1} & \cdots & x_r \end{pmatrix} \qquad \left(r = \frac{n(n+1)}{2} \right)$$

Define

$$\phi(x) = \max_{1 \le i \le N} \lambda_{\max} \left(A_i^{\mathrm{T}} P + P A_i \right) = \max_{1 \le i \le N, \|u\|=1} u^{\mathrm{T}} \left(A_i^{\mathrm{T}} P + P A_i \right) u. \tag{4}$$

If there exists x_* such that $P(x_*) > 0$ and $\phi(x_*) < 0$ then the matrix $P(x_*)$ is required solution. This problem can be reduced to the minimization of a convex function under convex constraints.

Consider the following convex minimization problem

$\phi(x) \to$ minimize.

$$\min_{\|v\|=1} v^{\mathrm{T}} P(x) v > 0 \tag{5}$$

Let $X \subset \mathbb{R}^n$ be a convex set and $F: X \to \mathbb{R}$ be convex function. The vector $g \in \mathbb{R}^n$ is said to be a subgradient of $F(x)$ at $x_* \in X$ if for all $x \in X$

$$F(x) \geq F(x_*) + g^{\mathrm{T}}(x - x_*).$$

The set of all subgradients of $F(x)$ at $x = x_*$ is denoted by $\partial F(x_*)$. If x_* is an interior point of X then the set $\partial F(x_*)$ is nonempty and convex. The following proposition follows from nondifferentiable optimization theory.

Proposition 1: Let $\phi(x)$ be defined as

$$\phi(x) = \max_{y \in Y} f(x, y) \tag{6}$$

where Y is compact, f (x, y) is continuous and differentiable in x. Then

$$\partial \phi(x) = \mathrm{conv}\left\{ \frac{\partial f(x, y)}{\partial x} : y \in Y(x) \right\}$$

where Y(x) is the set of all maximizing elements y in (6), i.e.

$$Y(x) = \{ y \in Y : f(x, y) = \phi(x) \}.$$

If for a given x the maximizing element is unique, i.e. $Y(x) = \{y(x)\}$ then $\phi(x)$ is differentiable at x and its gradient is

$$\nabla \phi(x) = \frac{\partial f(x, y)}{\partial x}.$$

In the case of the Function (4)

$$\partial\phi(x) = \operatorname{conv}\left\{\frac{\partial}{\partial x}\left(u^{\mathrm{T}}\left(A_i^{\mathrm{T}}P + PA_i\right)u\right) : i \text{ maximizes} \lambda_{\max}\left(A_i^{\mathrm{T}}P + PA_i\right),\right.$$

$$\left. u \text{ is a corresponding unit eigenvector}\right\}.$$

If for the given x the maximizing i is unique and $\lambda_{\max}\left(A_i^{\mathrm{T}}P + PA_i\right)$ is a simple eigenvalues, the differentiability of ϕ at the point x is guaranteed [13].

We investigate problem (5) by Kelley's cutting-plane method. This method converts the problem (5) to the problem

$$c^{\mathrm{T}}z \to \min$$
$$c_1(z) \geq 0, c_2(z) \geq 0, \ -1 \leq x_i \leq 1 \ (i = 1, 2, \cdots, r) \tag{7}$$

where

$$z = (x_1, x_2, \ldots, x_r, L)^{\mathrm{T}}, c = (0, 0, \ldots, 0, 1)^{\mathrm{T}}, c_1(z) = L - \phi(x), c_2(z) = \min_{\|v\|=1} v^{\mathrm{T}}Pv.$$

Let z^0 be a starting point and $z^0, z^1, \ldots, z^k$ be k+1 distinct points.

At the (k+1) th iteration, the cutting-plane algorithm solves the following LP problem

minimize $\qquad\qquad L$

subject to $\quad -h_1^{\mathrm{T}}\left(z^0\right)z \geq -h_1^{\mathrm{T}}\left(z^0\right)z^0 - c_1\left(z^0\right)$

$\qquad\qquad -h_2^{\mathrm{T}}\left(z^0\right)z \geq -h_2^{\mathrm{T}}\left(z^0\right)z^0 - c_2\left(z^0\right)$

$\qquad\qquad\qquad\qquad \vdots$

$\qquad\qquad -h_1^{\mathrm{T}}\left(z^k\right)z \geq -h_1^{\mathrm{T}}\left(z^k\right)z^k - c_1\left(z^k\right)$

$\qquad\qquad -h_2^{\mathrm{T}}\left(z^k\right)z \geq -h_2^{\mathrm{T}}\left(z^k\right)z^k - c_2\left(z^k\right)$

$\qquad\qquad\qquad -1 \leq x_i \leq 1 \tag{8}$

where $h_j(z^i)$ denotes a subgradient of $-c_j(z)$ at $z^i (i = 1,2)$.

Let z_*^k be the minimizer of the problem (8).

If z_*^k satisfies the inequality $\min\{c_1(z_*^k), c_2(z_*^k)\} \geq -\varepsilon$, where ε is a tolerance then z_*^k is an approximate solution of the problem (7). Otherwise define j^* as the index for the most negative $c_j(z_*^k)$, update the constraints in (8) by including the linear constraint

$$c_{j^*}\left(z^{k+1}\right) - h_{j^*}^T\left(z^{k+1}\right)\left(z - z^{k+1}\right) \geq 0$$

and repeat the procedure.

Recall that our aim is to find x_* such that $P(x_*) > 0$ and $\phi(x_*) < 0$, but not the solution of the minimization problem (5), (7).

Theorem 2: If there exists k such that

$$c_1\left(z_*^k\right) > L^k, c_2\left(z_*^k\right) > 0$$

where $z_*^k = \left(x_*^k, L^k\right)$ is the minimizer of the problem (8), then the matrix $P = P\left(x_*^k\right)$ is a common solution to (3).

Proof:

$$\phi\left(x_*^k\right) = L^k - c_1\left(z_*^k\right) < 0,$$

$$0 < c_2\left(z_*^k\right) = \min_{\|v\|=1} v^T P\left(x_*^k\right) v$$

and by (5), $P\left(x_*^k\right) > 0$ is a common solution to (3). For the problem (5), (7) Kelley's method gives the following.

Algorithm 1

Step 1: Take an initial point $z^0 = (x^0, L^0)^T$. Compute $\phi(x^0)$ and $c_2(z^0)$. If $\phi(x^0) < 0$ and $c_2(z^0) > 0$ stop; otherwise continue.

Step 2: Determine z_*^k by solving LP problem in (8). If $c_1(z_*^k) > L^k$ and $c_2(z_*^k) > 0$ then stop; otherwise continue. Set $z^{k+1} = z_*^k$, update the constraints in (8) and repeat the procedure.

Example 1: Consider the switched system

$$\dot{x} \in \{A_1, A_2, A_3\} x$$

where

$$A_1 = \begin{pmatrix} -2 & 5 & -6 \\ 0 & -8 & 0 \\ -5 & -2 & -20 \end{pmatrix}, \quad A_2 = \begin{pmatrix} -8 & 17 & -27 \\ 9 & -44 & 27 \\ 22 & -41 & -2 \end{pmatrix} \text{ and } A_3 = \begin{pmatrix} 4 & 9 & -2 \\ -6 & -8 & 4 \\ 1 & -10 & -6 \end{pmatrix}$$

are Hurwitz stable matrices.

Choose the initial point $z^0 = (x_1^0, x_2^0, x_3^0, x_4^0, x_5^0, x_6^0, L^0)^T = (1, 0, 0, 1, 0, 1, 1)^T$, then

$$P(x^0) = \begin{pmatrix} 1 & 0 & 0 \\ 0 & 1 & 0 \\ 0 & 0 & 1 \end{pmatrix},$$

$c_1(z^0) = -7.5247$, $c_2(z^0) = 1$ and $\phi(x^0) = \max\limits_{i \in \{1,2,3\}} \lambda_{max}\left(A_i^T P(x^0) + P(x_0) A_i\right) = 8.5247 > 0$.

We obtain $z^1 = (-1, 1, 1, 1, -1, 1, -27.9933)^T$ by solving LP problem in (8). Calculations give the following Table 1, and

$$z^{15} = (x^{15}, L^{15})^T = (0.7811, 0.6268, -0.1283, 1, -0.1254, 0.2383, -0.8206)^T.$$

Since $L^{15} - c_1(z^{15}) = -0.0287 < 0$ and $c_2(z^{15}) = -0.2075 > 0$,

Table 1: Kelley's algorithm for Example 1

k	Lk	C1(zk)	C2(zk)
1	−27.9933	−209.7383	−1.9999
2	−24.4038	−127.1153	−2.3326
3	−14.2596	−106.2473	−1.8092
4	−10.0497	−63.4433	−1.8878
⋮	⋮	⋮	⋮
14	−0.8465	−1.1881	0.2694
15	−0.8206	−0.7919	0.2075

$$P = P\left(x^{15}\right) = \begin{pmatrix} 0.7811 & 0.6268 & -0.1283 \\ 0.6268 & 1 & -0.1254 \\ -0.1283 & -0.1254 & 0.2383 \end{pmatrix}$$

is a common positive definite solution for

$$A_i^T P + P A_i < 0 \quad (i = 1, 2, 3).$$

STABLE MEMBER IN A POLYTOPE

This part is devoted to the following question: Given a matrix family $\{A(q) : q \in B\}$ where $B \subset \mathbb{R}^l$ is a compact, is there a stable matrix in this family?

In [12], a numerical algorithm has been proposed for a stable member in the affine matrix family $\{A(q) : q \in \mathbb{R}^l\}$. In this algorithm the uncertainty vector q varies in the whole space $\mathbb{R}^l$. On the other hand we consider the case where q varies in a box $B \subset \mathbb{R}^l$ and use the gradient algorithm for minimization of the nonconvex maximum eigenvalue function. By choosing appropriate step-size, we obtain the convergence.

Let $Z_1, Z_2, ..., Z_r \left(r = \dfrac{n(n+1)}{2} \right)$ be a basis for the subspace of $n \times n$ symmetric matrices and

$$Q_i(q) = (-Z_i) \oplus \left(A^T(q) Z_i + Z_i A(q) \right),$$

$$\phi(x, q) = \lambda_{\max} \left(\sum_{i=1}^{r} x_i Q_i(q) \right)$$

where $x = (x_1, x_2, ..., x_r)^T$, $q = (q_1, q_2, ..., q_k)^T$

Consider the problem

$$\phi(x, q) \rightarrow \text{minimize.}$$

$$\min_{\|x\|=1, q \in Q} v^T P(x) v > 0$$

Theorem 3: There is a stable matrix in the family $A(q)$ if and only if $\phi^* = \min_{(x,q)} \phi(x, q) < 0$.

Proof:

$\phi^* < 0 \Leftrightarrow$ there exists (x^*, q^*) such that $\sum_{i=1}^{r} x_i^* Q_i(q^*) < 0$

$$\Leftrightarrow \left(-\sum_{i=1}^{r} x_i^* Z_i \right) \oplus \left[A^T(q^*) \left(-\sum_{i=1}^{r} x_i^* Z_i \right) + \left(-\sum_{i=1}^{r} x_i^* Z_i \right) A(q^*) \right] < 0$$

$$\Leftrightarrow P(x^*) = \sum_{i=1}^{r} x_i^* Z_i > 0 \text{ and } A(q^*)^T P(x^*) + P(x^*) A(q^*) < 0.$$

By Lyapunov theorem, the matrix $A(q^*)$ is stable.

Example 2: Consider the family of matrices

$$A(q) = A_0 + q_1 A_1 + q_2 A_2 + q_3 A_3, \quad q_1, q_2, q_3 \in [-1, 1]$$

where

$$A_0 = \begin{pmatrix} 1 & 0 & 2 & 0 \\ -2 & 0 & -3 & 0 \\ -5 & 1 & -1 & 0 \\ -3 & -1 & 0 & -2 \end{pmatrix}, A_1 = \begin{pmatrix} -2 & 0 & 3 & 0 \\ -1 & 0 & -3 & 2 \\ -3 & 3 & -1 & 0 \\ -4 & -1 & 0 & -2 \end{pmatrix}, A_2 = \begin{pmatrix} -1 & 0 & 2 & 0 \\ -3 & -1 & -3 & 0 \\ -3 & 2 & -1 & 2 \\ -2 & -1 & 0 & -2 \end{pmatrix}, A_3 = \begin{pmatrix} -1 & 0 & 0 & 1 \\ -1 & -2 & 3 & -2 \\ 1 & -2 & 0 & -1 \\ 0 & -2 & 1 & -5 \end{pmatrix}.$$

For $q = (0,0,0)^T$, $A(q) = A_0$ is unstable. We apply the gradient algorithm to find a stable member in the family.

Let $x^0 = \left(\dfrac{1}{2}, 0, 0, 0, \dfrac{1}{2}, 0, 0, \dfrac{1}{2}, 0, \dfrac{1}{2}\right)^T$ and $q^0 = (1,0,0)^T$. So

$$a^0 = (x^0, q^0) = \left(\dfrac{1}{2}, 0, 0, 0, \dfrac{1}{2}, 0, 0, \dfrac{1}{2}, 0, \dfrac{1}{2}, 1, 0, 0\right)^T.$$

Then

$$\sum_{i=1}^{10} x_i^0 Q_i (q^0) = \begin{pmatrix} -P(x^0) & 0 \\ 0 & A(q^0)^T P(x^0) + P(x^0) A(q^0) \end{pmatrix}$$

$$= \begin{pmatrix} -1/2 & 0 & 0 & 0 & 0 & 0 & 0 & 0 \\ 0 & -1/2 & 0 & 0 & 0 & 0 & 0 & 0 \\ 0 & 0 & -1/2 & 0 & 0 & 0 & 0 & 0 \\ 0 & 0 & 0 & -1/2 & 0 & 0 & 0 & 0 \\ 0 & 0 & 0 & 0 & -1 & 0 & 5 & 0 \\ 0 & 0 & 0 & 0 & -3 & 0 & -6 & 2 \\ 0 & 0 & 0 & 0 & -8 & 4 & -2 & 0 \\ 0 & 0 & 0 & 0 & -7 & -2 & 0 & -4 \end{pmatrix}.$$

Maximum eigenvalue of this matrix and its corresponding unit eigenvector are

$\lambda_{\max} = 2.1866, \ v = \left(0,0,0,0,0.7644,-0.4480,-0.1668,-0.4324\right)^{\mathsf{T}}$

respectively. Gradient of the function Φ at a^0 is

$\nabla\phi\big|_{a^0} = \left(-2.44,-1.86,-11.04,-2.78,1.93,7.50,4.30,2.52,7.46,2.35,0.28,0.50,-2.73\right)^{\mathsf{T}}$.

The first tencomponent of the vector $a^1 = a^0 = t \cdot \nabla\phi\big|_{a^0}$ should be on the ten dimensional unit sphere. Therefore $t = 0.01531$ and

$a^1 = \left(0.53,0.02,0.16,0.04,0.47,-0.11,-0.06,0.46,-0.11,0.46,0.99,-0.007,0.04\right)^{\mathsf{T}}$.

After 4 steps, we get

$a^4 = \left(x^4,q^4\right) = \left(0.59,0.03,0.04,0.009,0.41,-0.05,-0.04,0.49,-0.15,0.45,0.98,-0.03,0.08\right)^{\mathsf{T}}$

and $\phi(x^4,q^4) = -0.2585 < 0$ therefore $A(q^4)$ is stable.

CONCLUSIONS

Two important problems from control theory are considered: the existence of common quadratic Lyapunov functions for switched linear systems and the existence of a stable member in a matrix polytope. We obtain new conditions which give new effective computational algorithms.

REFERENCES

1. Boyd, S. and Yang, Q. (1989) Structured and Simultaneous Lyapunov Functions for System Stability Problems. International Journal of Control, 49, 2215-2240. http://dx.doi.org/10.1080/00207178908559769
2. Büyükköroğlu, T., Esen, Ö. and Dzhafarov, V. (2011) Common Lyapunov Functions for Some Special Classes of Stable Systems. IEEE Transactions on Automatic Control, 56, 1963-1967. http://dx.doi.org/10.1109/tac.2011.2137510
3. Cheng, D., Guo, L. and Huang, J. (2003) On Quadratic Lyapunov Functions. IEEE Transactions on Automatic Control, 48, 885-890.http://dx.doi.org/10.1109/tac.2003.811274

4. Dayawansa, W.P. and Martin, C.F. (1999) A Converse Lyapunov Theorem for a Class of Dynamical Systems Which Undergo Switching. IEEE Transactions on Automatic Control, 44, 751-760. http://dx.doi.org/10.1109/9.754812

5. King, C. and Shorten, R. (2004) A Singularity Test for the Existence of Common Quadratic Lyapunov Functions for Pairs of Stable LTI Systems. Proceedings of the American Control Conference, Boston, 30 June-2 July 2004, 3881-3884.

6. Mason, O. and Shorten, R. (2006) On the Simultaneous Diagonal Stability of a Pair of Positive Linear Systems. Linear Algebra and Its Applications, 413, 13-23. http://dx.doi.org/10.1016/j.laa.2005.07.019

7. Narendra, K.S. and Balakrishnan, J. (1994) A Common Lyapunov Function for Stable LTI Systems with Commuting A-Matrices. IEEE Transactions on Automatic Control, 39, 2469-2471. http://dx.doi.org/10.1109/9.362846

8. Shorten, R.N. and Narendra, K.S. (2002) Necessary and Sufficient Conditions for the Existence of a Common Quadratic Lyapunov Function for a Finite Number of Stable Second Order Linear Time-Invariant Systems. International Journal of Adaptive Control and Signal Processing, 16, 709-728. http://dx.doi.org/10.1002/acs.719

9. Shorten, R.N., Mason, O., Cairbre, F.O. and Curran, P. (2004) A Unifying Framework for the SISO Circle Criterion and Other Quadratic Stability Criteria. International Journal of Control, 77, 1-8. http://dx.doi.org/10.1080/0020717031000 1633321

10. Liberzon, D. and Tempo, R. (2004) Common Lyapunov Functions and Gradient Algorithms. IEEE Transactions on Automatic Control, 49, 990-994.http://dx.doi.org/10.1109/tac.2004.829632

11. Polyak, B.T. and Shcherbakov, P.S. (2005) Hard Problems in Linear Control Theory: Possible Approaches to Solution. Automation and Remote Control, 66, 681-718.http://dx.doi.org/10.1007/s10513-005-0115-0

12. Polyak, B.T. and Shcherbakov, P.S. (1999) Numerical Search of Stable or Unstable Element in Matrix or Polynomial Families: A Unified Approach to Robustness Analysis and Stabilization. Robustness in Identification and Control Lecture Notes in Control and Information Sciences, 245, 344-358. http://dx.doi.org/10.1007/bfb0109879

13. Horn, R.A. and Johnson, C.R. (1985) Matrix Analysis. Cambridge University Press, Cambridge.

CITATION

Yılmaz, Ş. and Büyükköroğlu, T. (2014) On Two Problems for Matrix Polytopes. Applied Mathematics, 5, 2650-2656. doi: 10.4236/am.2014.517253.

An Inverse Problem of Matrix with Fixed Row and Column Sums and Its Application

Lijie Guo, ZhenJiang, and Shuo Zhou

Department of Mathematics, College of
Science, Northeast Dianli University, Jilin
City, 132012, P.R.China

ABSTRACT

In this paper, we study the inverse problem of an n×n matrix with fixed
row and column sums by using the singular value decomposition (SVD)
of a matrix. The least-square solutions of an n×n matrix with fixed row
and column sums are studied. The necessary and sufficient conditions
for the existence of the symmetric solutions and the expressions for
the inverse problem of a matrix AX=B are established. In addition, the
problem of using matrix with fixed row and column sums to construct
the optimal approximation to a given matrix is discussed, and the expression of the solution is provided. The algorithms are proposed and
applications to the theory of electric net are illustrated by examples.

INTRODUCTION AND MOTIVATION

Research at present, see [1-6]. In this paper, the least-square solutions
to the inverse problem of matrix with fixed row and column sums are
studied and its application is presented.

The notations used in this paper can be summarized as follows: $R^{n \times m}$
represents the set of all n×m matrices, $OR^{n \times n}$ denotes the set of all n×n
orthogonal matrices. Let R(A), N(A), A^+ and $||A||$ denote column space,
null space, the Moore-Penrose pseudo-inverse and the Frobenius norm

of A, respectively. The identity matrix of order n is written as I_n. $A \otimes B$ represents Kronecker product of) $A = (\alpha_{ij}) \in R^{n \times m}$ and $B \in R^{p \times q}$, i.e $A \otimes B = (\alpha_{ij}B)_{np \times mq}$ vec(A) denotes vectorization of A by column. For all A and B $R^{n \times m}$ their inner product is defined by $(A,B) = trB^T A = \sum_{i=1}^{n}\sum_{j=1}^{m}\alpha_{ij}b_{ij}$ then $R^{n \times m}$ is a Hilbert space. The norm of a matrix generated by the inner product is the Frobenius norm.

Definition: A matrix $A = (\alpha_{ij}) \in R^{n \times m}$ is said to be a fixed row and column sums matrix if $\sum_{k=1}^{n}\alpha_{ik} = \alpha$ $\qquad \alpha \in R,,$ for $i, j = 1, 2, ..., n$.

We denote the set of all n×n fixed row and column sums matrices by FSR$^{n \times m}$ An n×n matrix A is called symmetric and fixed row and column sums matrix, if $A \in$ FSR$^{n \times m}$ in addition, A is also a symmetric matrix. We denote the set of all n×n symmetric and fixed row and column sums matrices by SFSR$^{n \times n}$.

A Is a generalized doubly stochastic matrix if a=1. The generalized doubly stochastic matrices are widely used in the Markov stochastic processes.

A Is a doubly center matrix (indefinite admittance matrix) if a=0. The center and doubly center matrices are widely used in the theory of electric net, see [7].

Suppose that there are n terminals in a net. Inputted voltages and corresponding electric currents are written as $u_1, u_2, \cdots, u_n$ and $i_1, i_2, \cdots, i_n$, respectively. So electric current vector is $i = (i_1, i_2, \cdots, i_n)^T$ and voltage vector is $u = (u_1, u_2, \cdots, u_n)^T$ Their relations can be linearly expressed as $i = Au$, Based on the law of Kirchoff electric current and the relative law of electromotive force, A should satisfy the following constraints.

$Ae_n = 0$ and $e_n^T A = 0, e_n = (1, 1, \cdots, 1)^T \in R^n$

Problem 1: Given $X, B \in R^{n \times m}$ find an n×n fixed row and column sums matrix A such that

$$\|AX - B\| = \min \qquad\qquad (1)$$

Problem 2: Given $X, B \in R^{n \times m}$ find an $n \times n$ symme-tric and fixed row and column sums matrix A such that

$$AX = B \qquad (2)$$

Problem 3: Given $\tilde{A} \in R^{n \times m}$ find $\hat{A} \in S_E$ such that

$$\left\| \tilde{A} - \hat{A} \right\| = \inf_{A \in S_E} \left\| \tilde{A} - A \right\| \qquad (3)$$

Where SE denotes the solution set of problem 1. The column of matrix X is electric current vector by $i = (i_1, i_2 ..., i_n)^T$, and column of matrix B is voltage vector by $u = (u_1, u_2 ..., u_n)^T$ in the theory of electric net.

In this paper, the necessary and sufficient conditions under which S_E is nonempty are presented and the general form of S_E is given. Furthermore, the expression of the solution of problem 3 is provided.

The paper is organized as follows. In section II, the structures of $FSR^{n \times n}$ and $SFSR^{n \times n}$ are discussed, solvability conditions for problem 1 are established, and the general expression of these solutions in the real number field is provided. In section III, symmetric solvability conditions and the general expression of it for problem 2 are established. Base on the first three sections, in section IV, we prove the existence and uniqueness of the solution of problem 3 if the solution S_E of problem 1 is nonempty and establish the expression of this unique solution. In the last section, we introduce two algorithms for solving this kind of problems and test them through two numerical examples.

THE SOLVABILITY CONDITION AND GENERAL SOLUTIONS TO PROBLEM 1

First we discuss the structures of $FSRn^{\times n}$ and $SFSR^{n \times n}$.

Lemma 1: Given $e_n = (1, 1, ..., 1)^T \in R^n, \alpha \in R, A \in FSR^{n \times n}$ if and only if

$$Ae_n = ae_n \text{ and } e_n^T A = ae_n^T \qquad (4)$$

Lemma 2[8]: A necessary and sufficient condition for the pair of matrix equations $Ax = B$ and $Xc = D$, where $A \in R^{s \times n}$ $B \in R^{s \times n}$ $C \in R^{m \times l}$ and In $D \in R^{n \times l}$ to have a common solution is

$$AA^+B = B, DC^+C = D \text{ and } AD = BC \tag{5}$$

In this case, the general common solution is

$$X = A^+B + DC^+ - A^+ADC^+ + (I_n - A^+A)Y(I_m - CC^+) \tag{6}$$

where $Y \in R^{n \times m}$ is an arbitrary matrix.

Lemma 3: Given $e_n = (1,1,...,1)^T \in R^n,$.

$$J_n = \frac{1}{n}e_ne_n^T \qquad K_n = I_n - J_n$$

Then

$$e_n^+ = \frac{1}{n}e_n^T \qquad K_n^+ = K_n \qquad K_n^2 = K_n \qquad J_n^2 = J_n$$

And

$$K_nJ_n = J_nK_n = 0 \tag{7}$$

Theorem 1: Let $A \in DCR^{n \times n}$

$$e_n = (1,1,\cdots,1)^T \in R^n, \quad J_n = \frac{1}{n}e_ne_n^T$$

$$K_n = I_n - J_n$$

Then

$$A = aJ_n + K_nYK_n \qquad \text{for all } Y \in R^{n \times n} \tag{8}$$

In particular, A is a fixed row and column sums matrix, if Y is a symmetric matrix.

Proof: By lemma 1, $A \in FSR^{n \times m}$ is equivalent to $Ae_n = \alpha e_n$ and $e_n^T A = \alpha e_n^T$.

By lemma 2, $Ae_n = \alpha e_n$ and $e_n^T A = \alpha e_n^T$ have a common solution that can be expresses as

$$A = \frac{a}{n} e_n e_n^T + (I_n - (e_n^T)^+ e_n^T) Y (I_n - e_n e_n^+),$$

$$(9)$$

Where $Y \in R^{n \times n}$ is an arbitrary matrix. By using lemma 3, we get (8). Especially, if $Y \in R^{n \times n}$ is a symmetric matrix, then

$$A^T = a J_n^T + K_n^T Y^T K_n^T = a J_n + K_n Y K_n = A.$$

This completes the proof of the theorem.

Remark: If $A \in FSR^{n \times n}$, then $e_n = (1, 1, ..., 1)^T$ is an eigenvector of A corresponding to the eigenvalue $\lambda = \alpha$ an eigenvector x of A corresponding to $\lambda \neq \alpha$ satisfies $a_n^T x = 0$.

Lemma 4: Given $Y \in R^{s \times n}, X \in R^{m \times l}$ and Is $Z \in R^{s \times l}$ then the general form of the set

$$S^* = \left\{ A \in R^{n \times m} \mid f(A) = \| YAX - Z \| = \min \right\}$$

$$(10)$$

Can be expressed as

$$A = Y^+ Z X^+ + G - Y^+ Y G X X^+, \text{ for all } G \in R^{n \times m}$$

$$(11)$$

When $f(A) = 0, S^*$ is a nonempty set if and only if

$$YY^+ Z X^+ X = Z$$

$$(12)$$

Proof: $\| YAX - Z \| = \min$ is equivalent to

$$\| (X^T \otimes Y) \text{vec}(A) - \text{vec}(Z) \| = \min,$$

We obtain the general solution vec (A)

$$= (X^T \otimes Y)^+ \text{vec}(Z) + [I - (X^T \otimes Y)^+ (X^T \otimes Y)] \text{vec}(G)$$

$$= ((X^+)^T \otimes Y^+) \, \text{vec}\,(C)$$

$$+ [I - ((X^+)^T \otimes Y^+)(X^T \otimes Y)] \, \text{vec}\,(G)$$

$$= \text{vec}\,[Y^+ Z X^+] + \text{vec}\,(G) - \text{vec}\,[Y^+ YGXX^+],$$

Then we can give the general solution of $\|YAX - Z\| = \min$,

$$A = Y^+ Z X^+ + G - Y^+ YGXX^+.$$

A necessary and sufficient condition of $\|(X^T \otimes Y)\text{vec}\,(A) - \text{vec}\,(Z)\| = \min = 0$ to have a solution is $(X^T \otimes Y)(X^T \otimes Y)^* \, \text{vec}\,(Z) = \text{vec}\,(Z)$,

Then

$$\text{vec}\,(Z) = (X^T \otimes Y)(X^T \otimes Y)^+ \, \text{vec}\,(Z)$$

$$= [(X^+ X)^T \otimes (YY^+)] \, \text{vec}\,(Z) = \text{vec}\,(YY^+ ZX^+ X).$$

I.e. $\|YAX - Z\| = \min$ has solutions if and only $YY^+ X = Z$.

Theorem 2: Given X and $B \in R^{n \times m}$,

$e_n = (1,1,\ldots,1)^T$, $J_n = \dfrac{1}{n} e_n e_n^T$, $K_n = I_n - J_n$ Then the general solution of problem 1 can be expressed as

$$A = aJ_n + K_n [BZ^+ + G(I_n - ZZ^+)]K_n \qquad \text{for all } G \in R^{n \times n} \tag{13}$$

Where

$$Z = K_n X. \tag{14}$$

Proof: By theorem 1, we get

$$A = \frac{a}{n} e_n e_n^T + (I_n - \frac{1}{n} e_n e_n^T) Y (I_n - \frac{1}{n} e_n e_n^T)$$

$$= aJ_n + K_n Y K_n$$

Hence $\|AX - B\| = \min$ is equivalent to

$$\|K_n Y K_n X - (B - aJ_n X)\| = \min . \tag{15}$$

Let $Z = K_n X$, by lemma 4, we get

$$Y = K_n BZ^+ + G - K_n^+ K_n GZZ^+ . \tag{16}$$

Substituting (16) into (8), and applying lemma 3, we get (13).

Corollary 1: Given X and $B \in R^{n \times m}$,

$e_n = (1,1,...,1)^T$ Define $Z = K_n X$ Then

$\|AX - B\| = \min = 0$ has a solution if and only if

$$K_n BZ^+ Z = K_n B \qquad e_n^T B = a e_n^T X . \tag{17}$$

The general form of the solution can be expressed as

$$A = aJ_n + (B - aJ_n X)Z^+ K_n + K_n G(I_n - ZZ^+)K_n \tag{18}$$

Proof: Using a similar proof to Theorem 2 and applying lemma 4, we can prove this corollary.

Corollary 2: Given X and $B \in R^{n \times m}$,

$e_n = (1,1,...,1)^T$ if $e_n^T B = \alpha e_n^T X$, Then the general solution of problem 1 can be expressed as (18).

Proof: For (13), using $e_n^T B = \alpha e_n^T X$, we can prove this corollary.

Corollary 3: Given X and $B \in R^{n \times m}$,

$e_n = (1,1,...,1)^T$, if $e_n^T B = \alpha e_n^T X$, Then

$\|AX - B\| = \min = 0$ Has a solution if and only if

$$K_n BZ^+ Z = K_n B \tag{19}$$

Moreover, the general form of the solution can be expressed as (18).

Proof: Using Corollary 1, for $e_n^T B = \alpha e_n^T X$, we obtain $\|AX - B = \min = 0\|$ AX B min 0 has a solution if and only if (19).

SOLUTION TO PROBLEM 2

Lemma 5[9]: Given X and $B \in R^{n \times m}$, and

$\text{rank}(X = r,)$ rank then AX=B has a symmetric solution if and only if

$$X^T B = B^T X \text{ and } BX^+ X = B.$$

(20)

In addition, the general solution can be expressed as

$$A = BX^+ + (BX^+)^T (I_n - XX^+) + U_2 GU_2^T,$$

$$\text{for all } G \in SR^{(n-r) \times (n-r)}$$

(21)

Where the singular value decomposition (SVD) of X is

$$X = U \begin{pmatrix} \Sigma_0 \\ 0\ 0 \end{pmatrix} V^T = U_i \Sigma V_1^T$$

$$U = (U_1, U_2) \in R^{n \times n}, \ U_2 \in R^{n \times (n-r)} \quad U_2^T U_2 = I_{n-r} \text{ and } N(X^T) = R(U_2).$$

Theorem 3: Given X and $B \in R^{n \times m}$,

$(X, e_n) \in R^{n \times (m+1)}$ And $\text{rank}(X, e_n) = r$ then $AX \quad B$ has a symmetric solution satisfying $e_n^T A = \alpha e_n^T$ if and only if

$$X^T B = B^T X \quad e_n^T B = a e_n^T X \text{ and}$$

$$(B, ae_n)(X, e_n)^+ (X, e_n) = (B, ae_n)$$

(22)

The general solution can be expressed as

$$A = (B, ae_n)(X, e_n)^+$$

$$+ [(B, ae_n)(X, e_n)^+]^T [I_n - (X, e_n)(X, e_n)^+] + U_2 GU_2^T$$

$$\text{for all } G \in SR^{(n-r) \times (n-r)}$$

(23)

Where $U_2 \in R^{n \times (n-r)}$ $U_2^T U_2 = I_{n-r}$ and

$$N\begin{pmatrix} X^T \\ e_n^T \end{pmatrix} = R(U_2)$$

Proof: Using AX=B and $e_n^T A = \alpha e_n^T$ gives

$$A(X, e_n) = (B, \alpha e_n) \tag{24}$$

Applying lemma 5, we see that (24) is equivalent to (22), and the general solution can be expressed as (23).

SOLUTION TO PROBLEM 3

When the solution set S_E of problem 1 is nonempty, then S_E is a closed convex set. Therefore, the corresponding problem 3 has a unique optimal approximate solution, see [10]. We give the form of $\hat{A}$.

Theorem 4: Given $\tilde{A} \in R^{n \times n}, X, B \in R^{n \times m}$ and

$e_n = (1,1,...,1)^T \in R^n$ Suppose that the SVD of $K_n = I_n - \frac{1}{n} e_n e_n^T$ is given by.

$$K_n = I_n - \frac{1}{n} e_n e_n^T = U \begin{bmatrix} I_{n-1} & \\ & 0 \end{bmatrix} U^T = U_1 U_1^T$$

$$U = (U_1, U_2) \in OR^{n \times n} \text{ and } U_1^T U_1 = I_{n-1} \tag{25}$$

Then problem 3 has a unique optimal approximate solution which can be expressed as

$$\hat{A} = aJ_n + U_1 U_1^T BZ^+ U_1 U_1^T + U_1 U_1^T (\tilde{A} - aJ_n - U_1 U_1^T BZ^+ U_1 U_1^T) U_1 U_1^T \tag{26}$$

where Z=K$_n$X

Proof: Let

$$U^T(\tilde{A} - \frac{a}{n}e_n e_n^T - (I_n - \frac{1}{n}e_n e_n^T)BZ^+(I_n - \frac{1}{n}e_n e_n^T))U = \begin{bmatrix} \tilde{A}_{11} & \tilde{A}_{12} \\ \tilde{A}_{21} & \tilde{A}_{22} \end{bmatrix} \tag{27}$$

Where

$$\tilde{A}_{11} = U_1^T(\tilde{A} - \frac{a}{n}e_n e_n^T - U_1 U_1^T BZ^+ U_1 U_1^T)U_1 \text{ and}$$

$$\tilde{A}_{11} \in R^{(n-1)\times(n-1)}. \tag{28}$$

For all $A \in S_E$, by theorem 2, A can be expressed as

$$A = \frac{a}{n}e_n e_n^T + (I_n - \frac{1}{n}e_n e_n^T)[BZ^+ + G(I_n - ZZ^+)](I_n - \frac{1}{n}e_n e_n^T) = aJ_n + U_1 U_1^T[BZ^+ + G(I_n - ZZ^+)]U_1 U_1^T$$

for all $G \in R^{n\times n}$

$$\tag{29}$$

Hence Hence $\left\|\tilde{A} - A\right\|^2$

$$= \left\|\tilde{A}_{11} - U_1^T G(I_n - ZZ^+)U_1\right\|^2 + \left\|\tilde{A}_{12}\right\|^2 + \left\|\tilde{A}_{21}\right\|^2 + \left\|\tilde{A}_{22}\right\|^2,$$

For all $G \in R^{n\times n}$

Then $\left\|\tilde{A} - A\right\| = \min$ is equivalent to

$$\left\|\tilde{A}_{11} - U_1^T G(I_n - ZZ^+)U_1\right\| = \min. \tag{30}$$

By lemma 4, we get

$$U_1^T G(I_n - ZZ^+)U_1 = \tilde{A}_{11} \tag{31}$$

Substituting (31) into (13), and using (25), we obtain (26).

Corollary 5: Given $\tilde{A} \in R^{n \times n}, X$ and $B \in R^{n \times n}, e_n = (1,1,...,1)^T$ and $e_n^T B = \alpha e_n^T X$ let the SVD of $I_n - \frac{1}{n}e_1 n e_n^T$ be (25), then problem 3 has a unique optimal approximate solution which can be expressed as.

$$\hat{A} = aJ_n + (B - aJ_n X)Z^+ K_n + K_n(\tilde{A} - aJ_n - K_n BZ^+ K_n)K_n$$

(32)

Where $Z = K_n X$

After completing testing of a net, $\tilde{A}$ isn't a doubly center matrix. Applying Theorem 4, we can use the fixed row and column sums matrix to construct the optimal approximation to the matrix $\tilde{A}$.

ALGORITHM AND NUMERICAL EXAMPLES

Algorithm 1: (A matrix inverse problem) INPUT

1. Input known matrix $X, B \in R^{n \times m} (m \le n), \alpha \in R$.

2. input column vector $e_n = (1,1,...,1)^T \in R$. OUTPUT

Fixed row and column sums matrix $A \in R^{n \times n}$. THE ALGORITHM

1. compute $K_n = \left(I_n - \frac{1}{n}e_n e_n^T\right), Z = K_n X$.

2. Compute Z^+ if $K_n BZ^+ Z = K_n B, e_n^T B = \alpha e_n^T X$, next, otherwise, no solution, go to (4).

3. using (13), compute $A = \alpha J_n + K_n BZH^+ K_n$ (least square solution of minimum norm).

Algorithm 2: (optimal approximation) INPUT

1. input known matrix $X, B \in R^{n \times m}$ $m \le n$.

2. input column vector $e_n = (1,1,...,1)^T \in R^n$.

3. input known matrix $\tilde{A} \in R^{n \times n}, \alpha \in R$. OUTPUT.

Fixed row and column sums optimal approximation solution $\hat{A} \in R^{n \times n}$.
THE ALGORITHM

1. using (25),compute $U_1 \in R^{n \times (n-1)}$.

2. Compute $K_n = \left(I_n - \dfrac{1}{n} e_n e_n^T \right).Z = K_n X$.

3. Compute Z^+

4. using (26), compute $\hat{A} \in R^{n \times n}$.

Example 1: Given a=1

$$X = \begin{bmatrix} 1 & 2 & 9 \\ 2 & 2 & 5 \\ 1 & 1 & 3 \\ 2 & 3 & 2 \\ 2 & 5 & 8 \end{bmatrix} \quad B = \begin{bmatrix} 2.4000 & 5.6000 & 0.4000 \\ 1.0000 & 0.2000 & 7.8000 \\ 1.0000 & 3.2000 & 10.8000 \\ 4.0000 & 11.2000 & 7.8000 \\ -0.4000 & -7.2000 & 0.2000 \end{bmatrix}$$

$$\tilde{A} = \begin{bmatrix} -1.4000 & -0.4000 & 1.0100 & 0.4000 & 1.4000 \\ 0.8000 & 0.8000 & 0.2000 & -0.3900 & -0.4000 \\ 0.8000 & -0.2000 & 0.2000 & -0.4000 & 0.6100 \\ -1.2000 & -0.2000 & -0.8000 & 0.6000 & 2.6000 \\ 2.0000 & 1.0000 & 0.4000 & 0.8000 & -3.2000 \end{bmatrix}$$

Obviously, $\tilde{A}$ isn't a generalized doubly stochastic matrix. By algorithm 2, we correct the generalized doubly stochastic matrix and achieve

$$\hat{A} = \begin{bmatrix} -1.4018 & -0.4005 & 1.0029 & 0.3975 & 1.4019 \\ 0.8030 & 0.7989 & 0.1967 & -0.3962 & -0.4024 \\ 0.7985 & -0.1995 & 0.1984 & -0.4018 & 0.6012 \\ -1.1999 & -0.1994 & -0.8006 & 0.6003 & 2.5997 \\ 2.0002 & 1.0006 & 0.3994 & 0.8002 & -3.2004 \end{bmatrix}$$

Example 2: In an electric net design, the measure results of three groups of voltages and electric currents are given in Table 1. Find its indefinite admittance matrix.

Table1: The Measure Results of Three Groups of Voltages and Electric Currents

Interface n	l	2	3	4	5
Ui	l	2	1	2	2
1	0.8000	-0.6000	-0.6000	2.4000	-2.0000
U2	2	2	1	3	5
	3.0000	-2.4000	0.6000	8.6000	-9.8000
U3	9	5	3	2	8
13	-5.0000	2.4000	5.4000	2.4000	-5.2000

Applying algorithm 1, we get the indefinite admittance matrix. (Least square solution of minimum norm)

$$A = \begin{bmatrix} -0.9714 & -0.9143 & 0.1714 & 0.9857 & 0.7286 \\ 0.6000 & 0.6000 & -0.0000 & -0.6000 & -0.6000 \\ 0.6571 & -0.4286 & -0.0571 & -0.5286 & 0.3571 \\ -1.0572 & -0.5714 & -1.3429 & 0.8286 & 2.1429 \\ 0.7714 & 1.3143 & 1.2286 & -0.6857 & -2.6286 \end{bmatrix}$$

After completing testing of a net, we get initial estimation to the indefinite admittance matrix

$$\tilde{A} = \begin{bmatrix} -1.6000 & -0.6000 & 0.8100 & 0.2000 & 1.2000 \\ 0.6000 & 0.6000 & -0.0000 & -0.5900 & -0.6000 \\ 0.6000 & -0.4000 & -0.0000 & -0.6000 & 0.4100 \\ -1.4000 & -0.4000 & -1.0000 & 0.4000 & 2.4000 \\ 1.8000 & 0.8000 & 0.2000 & 0.6000 & -3.4000 \end{bmatrix},$$

Obviously, $\tilde{A}$ isn't a doubly center matrix. By algorithm 2, we correct the indefinite admittance matrix by using test data of table1 and achieve

$$\hat{A} = \begin{bmatrix} -1.6018 & -0.6005 & 0.8029 & 0.1975 & 1.2019 \\ 0.6030 & 0.5989 & -0.0033 & -0.5962 & -0.6024 \\ 0.5985 & -0.3995 & -0.0016 & -0.6018 & 0.4012 \\ -1.3999 & -0.3994 & -1.0006 & 0.4003 & 2.3997 \\ 1.8002 & 0.8006 & 0.1994 & 0.6002 & -3.4004 \end{bmatrix}$$

Note that $\tilde{A}$ is a doubly center matrix.

REFERENCES

1. G. M. L. Gladwell, Inverse problems in vibration, Martinus Nijhoff: Dordrecht, The Netherlands, Boston, MA, 1986. 1606 Lijie Guo et al. / Energy Procedia 17 (2012) 1598 – 1606
2. V. Barcilon, "Sufficient conditions for the solution of inverse problem for a vibrating beam," Inverse Problems, vol. 3, pp. 181–193, 1987.
3. A. Kirsch, An introduction to the mathematical theory of inverse problems, Springer-Verlag New York, 1996.
4. H. Dai, L. Peter, "Linear matrix equations from an inverse problem of vibration theory," Linear Algebra Appl. vol. 246, pp. 31-47, 1996.
5. Bassam Mourad, "An inverse problem for symmetric doubly stochastic matrices," Inverse Problems, vol. 19, pp. 821- 831, 2003.
6. Chung-Tsun Shieh, "Some inverse problems on Jacobi matrices," Inverse Problems, vol. 20, pp. 589-600, 2004.
7. Shuo Zhou, Baisheng Wu, "The inverse problem of doubly center matrices and its application to the theory of electric network," Chinese Journal of Engineering Mathematics, vol. 24, pp. 611-617, 2007.
8. Wang Songgui and Yang Zhenhai, Generalized Inverse Matrix and Its Application, Beijing: Beijing Industry Press in Chinese, 1996.
9. Zhang Lei, "Inverse eigenproblems for a class of symmetric matrices," Numer . Math. J. Chinese Univ., vol. 12, pp. 65- 71, 1990.
10. Zhang Lei, "Approximation in closed convex cones and its applications in numerical analysis," Annual of hunan mathematics in Chinese, vol. 6, pp. 43-48, 1986.

CITATION

Lijie Guo, Zhen Jiang, Shuo Zhou, An Inverse Problem of Matrix with Fixed Row and Column Sums and its Application, Energy Procedia, Volume 17, Part B, 2012, Pages 1598-1606, ISSN 1876-6102, http://dx.doi.org/10.1016/j.egypro.2012.02.287.

Expansion Formulas for the Inertias of Hermitian Matrix Polynomials and Matrix Pencils of Orthogonal Projectors

Yongge Tian
China Economics and Management Academy, Central University of Finance and Economics, Beijing 100081, China

7

ABSTRACT

This paper gives a group of expansion formulas for the inertias of Hermitian matrix polynomials $A-A^2$, $I-A^2$ and $A-A^3$ through some congruence transformations for block matrices, where A is a Hermitian matrix. Then, the paper derives various expansion formulas for the ranks and inertias of some matrix pencils generated from two or three orthogonal projectors and Hermitian unitary matrices. As applications, the paper establishes necessary and sufficient conditions for many matrix equalities to hold, as well as many inequalities in the Löwner partial ordering to hold.

INTRODUCTION

Throughout this paper, $\mathbb{C}^{m \times n}$ and $\mathbb{C}_H^m$ denote the collections of all $m \times n$ complex matrices and all $m \times m$ complex Hermitian matrices, respectively. The symbols $A*$, $r(A)$ and $\mathfrak{R}(A)$ stand for the conjugate transpose, rank, range (column space) of a matrix $A \in \mathbb{C}^{m \times n}$, respectively; I_m denotes the identity matrix of order m; $[A, B]$ denotes a row block matrix consisting of A and B. Two Hermitian matrices A and B of the same size are said to be congruent if there is an invertible matrix S such that $SAS* = B$. We write $A > 0$ ($A \geq 0$) if A is Hermitian positive (nonnegative) definite. Two Hermitian matrices A and B of the same size are said to satisfy the inequality A > B (A $\geq_{\mathbb{C}}$ B) in the Löwner partial ordering if A − B is positive (nonnegative) definite; cf. Löwner [35, p. 177], and also Marshall and Olkin [37, p. 462]. The Moore–Penrose inverse of A $\in \mathbb{C}^{m \times n}$, denoted by $A^\dagger$, is defined to be the unique matrix X $\in \mathbb{C}^{n \times m}$ satisfying the four matrix equations AXA = A, XAX = X, (A X) * = A X and (X A) * = X A. In particular, $a^\dagger = a^{-1}$ if a = 0 and $a^\dagger = 0$ if a = 0 for a scalar a $\in \mathbb{C}$. Some well-known equality for the Moore–Penrose inverse are given by see [62]. Results on the Moore–Penrose inverse can be found, e.g., in [12, 13, 29].

$$A^\dagger = A^*\left(AA^*\right)^\dagger = \left(A^*A\right)^\dagger A^* = A^*\left(A^*AA^*\right)^\dagger A^*, \qquad A^* = A^\dagger AA^* = A^*AA^\dagger;$$

(1)

A matrix A $\in \mathbb{C}^{m \times m}$ is called an orthogonal projector if it is both idempotent and Hermitian, i.e., $A^2 = A = A*$; the collection of all orthogonal projectors of order m is denoted by $\mathbb{C}^m_{Op}$. A matrix A $\in \mathbb{C}^{m \times m}$ is said to be Hermitian unitary if A = A* = A^{-1}, and the collection of all Hermitian unitary matrices of order m is denoted by $\mathbb{C}^m_{HU}$. A matrix X $\in \mathbb{C}^{m \times m}$ is called the orthogonal projector onto the range $\mathfrak{R}$ (A) of A $\in \mathbb{C}^{m \times m}$, denoted by X = P A, if it satisfies $\mathfrak{R}$ (X) = $\mathfrak{R}$ (A) and $X^2 = X = X*$. It can be seen from the definition of the Moore–Penrose inverse that the orthogonal projector onto $\mathfrak{R}$ (A) can uniquely be represented as $P_A =$ $AA^\dagger$. Further, denote $E_A = I_m - AA^\dagger$ and $F_A = I_n - A^\dagger A$, both of which are orthogonal projectors onto the null spaces of A* and A, respectively, and their ranks are given by $r(E_A) = m - r(A)$ and $r(F_A) = n - r(A)$.

When considering a Hermitian matrix, we are usually concerned with distributions of the eigenvalues of the matrix, as well as its definiteness. Recall that the eigenvalues of a Hermitian matrix A $\in \mathbb{C}^{m \times m}$ A $\in$ Cm H are all real numbers, and the inertia of A is defined to be the triplet

$$\text{In}(A) = \left\{i_+(A), i_-(A), i_0(A)\right\},$$

Where $i_+(A)$, $i_-(A)$ and $i_0(A)$ are the numbers of the positive, negative and zero eigenvalues of A counted with multiplicities, respectively. The two numbers i_+ (A) and i_- (A) are usually called the partial inertia of A; see, e.g., [9]. The difference i_+ (A) – i_- (A), denoted by s (A), is usually called the signature of A. For a matrix $A \in \mathbb{C}_H^m$, we have r (A) = i_+ (A) + i_- (A) and i_0 (A) = m – r (A). Hence, once the partial inertia $i_\pm$ (A) is determined, r (A), i_0 (A) and s (A) are obtained as well.

This paper aims at establishing some basic formulas for inertias of certain simple polynomials consisting of a Hermitian matrix, and then using the formulas to derive a variety of equalities for ranks/inertias of various matrix expressions consisting of orthogonal projectors. As applications, the author gives necessary and sufficient conditions for a wealth of matrix equalities and inequalities consisting of orthogonal projectors to hold.

Note that the inertia of a Hermitian matrix describes the sign distribution of the real eigenvalues of the matrix. Hence, it can be used to characterize definiteness of the matrix. The following results are obvious from the definitions of the rank/inertia of a matrix.

Lemma 1.1: Let $A \in \mathbb{C}^{m \times m}, B \in \mathbb{C}^{m \times m}$ and $C \in \mathbb{C}_H^m$ then

a) A is non-singular if and only if r (A) = m.
b) B = 0 if and only if r (B) = 0.
c) C > 0 (C < 0) if and only if i_+(C) = m (i_-(C) = m).
d) C ≥ 0 (C ≤ 0) if and only if i_-(C) = 0 (i_+(C) = 0).

This lemma shows that once certain formulas for ranks/inertias of Hermitian matrices and their operations are derived, we can use them to characterize equalities and inequalities for matrices. This basic algebraic method, which we refer to as the matrix rank/inertia method, is quite effective to solve various problems on conditional matrix equalities and inequalities in matrix theory and applications. It is well known in undergraduate linear algebra course that a direct method for computing the inertia of a Hermitian matrix is to reduce the matrix to a diagonal form by congruence transformations. This method is unstable for computing the exact inertia of a general matrix from the numerical point of view, so that computing the inertia of a matrix is regarded as a hard problem in linear algebra and no method is known to get the inertia of a general matrix exactly; see, e.g., [27,28]. From the symbolical point of view, the congruence transformation is the only method to study algebraic properties of Hermitian matrices. Without much effort, many closed-form formulas for ranks/inertias of Hermitian matrices and their operation can be established through congruence transformations; see the author's recent papers [50, 51, 54], and the results in the sections below

We shall repeatedly use the simple or well-known results on ranks and inertias of (Hermitian) matrices in the following lemmas.

Lemma 1.2: (See [36, Theorem 19].) Let $A \in \mathbb{C}^{m \times m}$, $B \in \mathbb{C}^{m \times k}$ and $C \in \mathbb{C}^{l \times n}$ are given. Then

$$r[A, B] = r(A) + r(E_A B) = r(B) + r(E_B A), \tag{1.2}$$

$$r\begin{bmatrix} A & B \\ C & 0 \end{bmatrix} = r(B) + r(C) + r(E_B A F_C). \tag{1.3}$$

In particular

$$r[A, B] = r(A) + r(B) \quad \Leftrightarrow \quad \mathscr{R}(A) \cap \mathscr{R}(B) = \{0\}. \tag{1.4}$$

Lemma 1.3: Let $A \in \mathbb{C}_H^m$, $B \in \mathbb{C}_H^n$, $Q \in \mathbb{C}^{m \times m}$, and assume that $P \in \mathbb{C}^{m \times m}$ is non-singular. Then,

$$i_\pm(PAP^*) = i_\pm(A), \tag{1.5}$$

$$i_\pm(A^{2k-1}) = i_\pm(A) \quad and \quad i_\pm(A^\dagger) = i_\pm(A) \quad for \ any \ integer \ k \geqslant 1 \tag{1.6}$$

$$i_\pm(\lambda A) = \begin{cases} i_\pm(A) & if \ \lambda > 0, \\ i_\mp(A) & if \ \lambda < 0, \end{cases} \tag{1.7}$$

$$i_\pm\begin{bmatrix} A & 0 \\ 0 & B \end{bmatrix} = i_\pm(A) + i_\pm(B), \tag{1.8}$$

$$i_\pm \begin{bmatrix} 0 & Q \\ Q^* & 0 \end{bmatrix} = r(Q).$$
(1.9)

Eq. (1.5) is the well-known Sylvester's law of inertia, which was first established in 1852 by Sylvester [45] (see also [30, Theorem 4.5.8] and [38, p. 377]). Eq. (1.6) follows from the fact that the signs of nonzero eigenvalues of A, A^{2k-1} and A^+ are the same. Eqs. (1.7) and (1.8) are obvious from the definition of inertia, and (1.9) is well known; see, e.g., [25, 26]. We also need the following results on ranks/inertias of Hermitian matrices

Lemma 1.4: Let A, B $\in \mathbb{C}_H^m$, and assume that A B = B A. Then,

a) Both A^{2k-1} B and A B^{2k-1} are Hermitian, and

$$i_\pm \left(A^{2k-1} B \right) = i_\pm \left(AB^{2k-1} \right) = i_\pm (AB)$$
(1.10)

For any integer $k \geq 1$

b) If $B \geq 0$, then $A^{2k} B \geq 0$, and

$$i_+ \left(A^{2k} B \right) = r \left(A^{2k} B \right) = r(AB)$$

For any integer $k \geq 1$

Proof: It is well known that under the conditions A = A∗, B = B∗ and A B = B A, there exists a unitary matrix U such that A = U diag$\{\lambda_1,...,\lambda_m\}$ U∗ and B = U diag$\{\mu_1,...,\mu_m\}$U∗, where $\lambda_1,...,\lambda_m$ and $\mu_1,...,\mu_m$ are the real eigenvalues of A and B, respectively. In this case,

$AB = U \text{ diag}\{\lambda_1\mu_1, ..., \lambda_m\mu_m\}U^*,$

$A^{2k-1}B = U \text{ diag}\{\lambda_1^{2k-1}\mu_1, ..., \lambda_m^{2k-1}\mu_m\}U^*, \qquad AB^{2k-1} = U \text{ diag}\{\lambda_1\mu_1^{2k-1}, ..., \lambda_m\mu_m^{2k-1}\}U^*.$

Where $\lambda_i\mu_i$, $\lambda_i^{2k-1}\mu_i$ and $\lambda_i\mu_i^{2k-1}$ have the same sign, i = 1... m. Hence, (1.10) follows. Under the condition $B \geq 0$, we have A^{2k} B = U diag$\{\lambda_1^{2k}\mu_1,...,\lambda_m^{2k}\mu_m\}$U∗ ≥ 0. Thus (1.11) follows.

Lemma 1.5: (See [50, Theorem 2.3].) Let $A \in \mathbb{C}_H^m$ Cm H, $B \in \mathbb{C}^{m\times m}$, $D \in \mathbb{C}_H^n$, and denote

$$M_1 = \begin{bmatrix} A & B \\ B^* & 0 \end{bmatrix}, \qquad M_2 = \begin{bmatrix} A & B \\ B^* & D \end{bmatrix}.$$

Then,

$$i_\pm(M_1) = r(B) + i_\pm(E_B A E_B), \tag{1.12}$$

$$r(M_1) = 2r(B) + r(E_B A E_B), \tag{1.13}$$

$$i_\pm(M_2) = i_\pm(A) + i_\pm \begin{bmatrix} 0 & E_A B \\ B^* E_A & D - B^* A^\dagger B \end{bmatrix}, \tag{1.14}$$

$$r(M_2) = r(A) + r \begin{bmatrix} 0 & E_A B \\ B^* E_A & D - B^* A^\dagger B \end{bmatrix}. \tag{1.15}$$

In particular

a) If $A \geq 0$, then
$$i_+(M_1) = r[A, B], \qquad i_-(M_1) = r(B), \qquad r(M_1) = r[A, B] + r(B). \tag{1.16}$$

b) If $A \leq 0$, then
$$i_+(M_1) = r(B), \qquad i_-(M_1) = r[A, B], \qquad r(M_1) = r[A, B] + r(B). \tag{1.17}$$

c) If $(B) \subseteq (A)$, then
$$i_\pm(M_2) = i_\pm(A) + i_\pm(D - B^* A^\dagger B), \qquad r(M_2) = r(A) + r(D - B^* A^\dagger B). \tag{1.18}$$

d) If $\mathcal{R}(B) \cap \mathcal{R}(A) = \{0\}$ and $\mathcal{R}(B*) \cap \mathcal{R}(D) = \{0\}$, then
$$i_\pm(M_2) = i_\pm(A) + i_\pm(D) + r(B), \qquad r(M_2) = r(A) + 2r(B) + r(D). \tag{1.19}$$

In order to simplify block matrices, we adopt the following three types of elementary block matrix operations (EBMOs, for short): (I) interchange two block rows (columns) in a block matrix; (II) multiply a block row (column) by a non-singular matrix from the left-hand (right-hand) side in a block matrix; (III) add a block row (column) multiplied by a matrix from the left-hand (right-hand) side to another block row (column). It is obvious that EBMOs don't change the rank of a block matrix.

The rest of this paper is organized as follows. In Section 2, we first construct a group of congruence transformations for some block matrices consisting of a Hermitian matrix A and its operations. From the congruence transformations and the Sylvester's law of inertia, we derive some basic formulas for the partial inertias of the matrix polynomials $A - A^2$, $I_m - A^2$ and $A - A^3$ in terms of the partial inertias of A and $I_m \pm A$, and present some variations of the expansion formulas. In Sections 3 and 4, we give a variety of expansion formulas for the partial inertias of matrix pencils generated from two or more orthogonal projectors and their operations, and present various consequences and applications of these formulas. Section 5 gives some expansion formulas for the inertias of orthogonal projectors onto the ranges of block matrices and sub matrices. Section 6 gives expansion formulas for partial inertias of some matrix pencils generated from Hermitian unitary matrices. Section 7 proposes some problems on inertias of Hermitian matrices for further consideration

EXPANSION FORMULAS FOR RANKS/INERTIAS OF SOME POLYNOMIALS OF A HERMITIAN MATRIX

When considering a quantity in mathematics, it is always desirable to establish some informative expansion formulas for the quantity. Once certain expansion formulas for the quantity are established, we can use them to derive various properties of the quantity. This is a quite inclusive but challenging topic in mathematics and applications. In matrix theory, many numerical characteristics of matrices can be defined, and of course, some valuable expansion formulas for such numerical characteristics are expected to establish. For a given Hermitian matrix, one of the most basic concepts associated with the matrix is its inertia. In a recent paper [50], the present author collected some well-known formulas for inertias of Hermitian matrices, and also showed many new expansion formulas for inertias of block Hermitian matrices, products of Hermitian matrices, and sums of Hermitian matrices. In particular, the present author gave some expansion formulas for the inertias of A ± B, where A and B are both Hermitian matrices. As a continuation, we derive expansion formulas for the ranks/inertias of the three matrix polynomials $A - A^2$, $I_m - A^2$ and $A - A^3$, where A is a Hermitian matrix of order m.

For any given square matrix A of order m, the following three simple and interesting rank formulas are well known in undergraduate linear algebra.

$$r(A - A^2) = r(A) + r(I_m - A) - m,$$

(2.1)

$$r(I_m - A^2) = r(I_m + A) + r(I_m - A) - m,$$

(2.2)

$$r(A - A^3) = r(A) + r(I_m + A) + r(I_m - A) - 2m$$

(2.3)

These rank formulas can be proved in about a paragraph by using only an idea of elementary matrix operations to certain block matrices consisting of I_m, A and their operations. For example, (2.1) can be derived from the following two-sided elementary block matrix operations

$$\begin{bmatrix} I_m & 0 \\ -A & I_m \end{bmatrix} \begin{bmatrix} I_m & I_m - A \\ A & 0 \end{bmatrix} \begin{bmatrix} I_m & -I_m + A \\ 0 & I_m \end{bmatrix} = \begin{bmatrix} I_m & 0 \\ 0 & A^2 - A \end{bmatrix},$$

(2.4)

$$\begin{bmatrix} I_m & -I_m \\ 0 & I_m \end{bmatrix} \begin{bmatrix} I_m & I_m - A \\ A & 0 \end{bmatrix} \begin{bmatrix} I_m & 0 \\ -I_m & I_m \end{bmatrix} = \begin{bmatrix} 0 & I_m - A \\ A & 0 \end{bmatrix}.$$

(2.5)

More elementary block matrix transformations and the corresponding rank equalities for matrix polynomials can be found in the literature; see, e.g., [1, 55, 57]. The three elementary formulas in (2.1)–(2.3) show such an interesting fact that the ranks of the three matrix polynomials $A - A^2$, $I_m - A^2$ and $A - A^3$ can be calculated through the algebraic operations of the ranks of their multiplication factors A and $I_m \pm A$. Hence, (2.1)–(2.3) can be called as expansion formulas for the ranks of the three matrix polynomials. Eqs. (2.1)– (2.3) can be used to characterize some basic algebraic properties of $A - A^2$, $I_m - A^2$ and $A - A^3$, such as, the no singularity of three matrix polynomials, as well as the idempotency, involution and tripotency of A.

The rank expansion formulas in (2.1)–(2.3) hold, of course, for any Hermitian matrix A of order m. Also, recall that the rank of a Hermitian matrix A is the sum of the partial inertia of A. Therefore, the rank and inertia of a Hermitian matrix should share the same mechanism. In fact, the expansion formulas in (1.12)–(1.15) show such reasonable separations between the ranks and partial inertias of block Hermitian matrices. This fact prompts us to consider some reasonable separations of (2.1)–(2.3) into certain expansion formulas for the partial inertias of $A- A^2$, $I_m - A^2$ and $A- A^3$. The approach pursued in this section is in a similar spirit to (2.4) and (2.5), but relies primarily on congruence transformations for some block Hermitian matrices consisting of I_m, A and their operations, as well as the formulas in Lemma 1.3.

Theorem 2.1: Let $A \in \mathbb{C}_H^m$. Then,

$$i_+\left(A - A^2\right) = i_+(A) + i_+(I_m - A) - m, \tag{2.6}$$

$$i_-\left(A - A^2\right) = i_-(A) + i_-(I_m - A), \tag{2.7}$$

$$s\left(A - A^2\right) = s(A) + s(I_m - A) - m, \tag{2.8}$$

$$i_+\left(I_m - A^2\right) = i_+(I_m + A) + i_+(I_m - A) - m, \tag{2.9}$$

$$i_-\left(I_m - A^2\right) = i_-(I_m + A) + i_-(I_m - A), \tag{2.10}$$

$$s\left(I_m - A^2\right) = s(I_m + A) + s(I_m - A) - m, \tag{2.11}$$

$$i_\pm\left(A - A^3\right) = i_\pm(A) + i_\mp(I_m + A) + i_\pm(I_m - A) - m, \tag{2.12}$$

$$s\left(A - A^3\right) = s(A) + s(I_m - A) - s(I_m + A). \tag{2.13}$$

Hence,

a) $A - A^2 > 0$ $(A - A^2 \geq 0)$ if and only if $I_m > A > 0$ $(I_m \geq A \geq 0)$, i.e., A is a strict contraction (contraction).

b) $A - A^2 < 0$ $(A - A^2 \leq 0)$ if and only if $i_-(I_m - A) + i_-(A) = m$ $(i_+(I_m - A) + i_+(A) = m)$.

c) $I_m - A2 > 0$ $(I_m - A2 \geq 0)$ if and only if $I_m > A > -Im$ $(I_m \geq A \geq -I_m)$.

d) $I_m - A^2 < 0$ $(I_m - A^2 \leq 0)$ if and only if $i_-(I_m + A) + i_-(I_m - A) = m$ $(i_+(I_m + A) + i_+(I_m - A) = m)$.

e) $A - A^3 > 0$ $(A - A^3 \geq 0)$ if and only if $i_-(I_m + A) + i_+(I_m - A) + i+(A) = 2m$ $(i_+(I_m + A) + i_-(I_m - A) + i_-(A) = m)$.

f) $A - A^3 < 0$ $(A - A^3 \leq 0)$ if and only if $i_+(I_m + A) + i_-(I_m - A) + i_-(A) = 2m$ $(i_-(I_m + A) + i_+(I_m - A) + i_+(A) = m)$.

Proof: It is easily seen from (1.5) that if

$$P_1 M P_1^* = N_1 \quad \text{and} \quad P_2 M P_2^* = N_2 \tag{2.14}$$

For three Hermitian matrices M, N_1 and N_2 of the same size, where P_1 and P_2 are both non-singular, then

$$i_\pm(M) = i_\pm(N_1) = i_\pm(N_2). \tag{2.15}$$

Now let

$$M_1 = \begin{bmatrix} 2^{-1}I_m & 2^{-1}I_m - A \\ 2^{-1}I_m - A & 2^{-1}I_m \end{bmatrix}, \quad M_2 = \begin{bmatrix} I_m & A \\ A & I_m \end{bmatrix}, \quad M_3 = \begin{bmatrix} -A & 0 & I_m \\ 0 & A & A \\ I_m & A & 0 \end{bmatrix}. \tag{2.16}$$

Then, the three block matrices are all Hermitian. Also, it is easily verified that

$$P_1 M_1 P_1^* = 2 \begin{bmatrix} I_m - A & 0 \\ 0 & A \end{bmatrix}, \quad P_1 = \begin{bmatrix} I_m & I_m \\ I_m & -I_m \end{bmatrix}, \tag{2.17}$$

$$Q_1 M_1 Q_1^* = \begin{bmatrix} 2^{-1}I_m & 0 \\ 0 & 2(A - A^2) \end{bmatrix}, \qquad Q_1 = \begin{bmatrix} I_m & 0 \\ -I_m + 2A & I_m \end{bmatrix}, \tag{2.18}$$

$$P_2 M_2 P_2^* = 2 \begin{bmatrix} I_m + A & 0 \\ 0 & I_m - A \end{bmatrix}, \qquad P_2 = \begin{bmatrix} I_m & I_m \\ I_m & -I_m \end{bmatrix}, \tag{2.19}$$

$$Q_2 M_2 Q_2^* = \begin{bmatrix} I_m & 0 \\ 0 & I_m - A^2 \end{bmatrix}, \qquad Q_2 = \begin{bmatrix} I_m & 0 \\ -A & I_m \end{bmatrix}, \tag{2.20}$$

$$P_3 M_3 P_3^* = \begin{bmatrix} I_m - A & 0 & 0 \\ 0 & A & 0 \\ 0 & 0 & -I_m - A \end{bmatrix}, \qquad P_3 = \begin{bmatrix} \frac{1}{\sqrt{2}}I_m & -\frac{1}{\sqrt{2}}I_m & \frac{1}{\sqrt{2}}I_m \\ 0 & I_m & 0 \\ \frac{1}{\sqrt{2}}I_m & \frac{1}{\sqrt{2}}I_m & -\frac{1}{\sqrt{2}}I_m \end{bmatrix}, \tag{2.21}$$

$$Q_3 M_3 Q_3^* = \begin{bmatrix} 0 & 0 & I_m \\ 0 & A - A^3 & 0 \\ I_m & 0 & 0 \end{bmatrix}, \qquad Q_3 = \begin{bmatrix} I_m & 0 & 2^{-1}A \\ -A & I_m & -A^2 \\ 0 & 0 & I_m \end{bmatrix}, \tag{2.22}$$

And that the six block matrices P_i and Q_i, $i = 1, 2, 3$, are all non-singular. Applying (2.15) to (2.17)–(2.22) and simplifying by (1.7)–(1.9), we obtain the following equalities for the partial inertias of M_1, M_2 and M_3

$$i_\pm(M_1) = i_\pm(A) + i_\pm(I_m - A) = i_\pm(I_m) + i_\pm(A - A^2),$$

$$i_\pm(M_2) = i_\pm(I_m + A) + i_\pm(I_m - A) = i_\pm(I_m) + i_\pm(I_m - A^2),$$

$$i_\pm(M_3) = i_\mp(I_m + A) + i_\pm(I_m - A) + i_\pm(A) = m + i_\pm(A - A^3).$$

Thus, we have (2.6)–(2.13). Applying Lemma 1.1 to (2.6), (2.7), (2.9), (2.10) and (2.12) yields (a)–(f).

Remark 2.2

a) Adding (2.6) and (2.7), (2.9) and (2.10), and the two equalities in (2.12) gives rise to the rank formulas in (2.1)–(2.3). In other words, the rank formulas in (2.1)–(2.3) can reasonably be separated into two groups of equalities for the partial inertias of $A - A^2$, $I_m - A^2$ and $A - A^3$. Namely, (2.6)–(2.13) are refinements of the rank formulas

in (2.1)–(2.3) for a Hermitian matrix. Similarly, we can do such reasonable separations for many rank formulas for Hermitian matrix
expressions; see [50] for more details

b) Notice that (2.6)–(2.13) are derived in about a paragraph by only
using the well-known Sylvester's law of inertia and some trivial results in (1.7)–(1.9). Also note that the formulas in (2.6)–(2.13) and
(2.1)–(2.3) are matched quite well. Hence, (2.6)–(2.13) and their
variations should be proved/published in an earlier period of linear
algebra, and thus become some classical contents on inertias of
Hermitian matrices in linear algebra. As demonstrated in the following several sections, many simple and valuable results on ranks/
inertias of Hermitian matrices can easily be proved by some elementary methods. This fact also shows that the theory of ranks/in-
ertias of matrices was not so sufficiently developed in the past cen-
turies that a huge amount of valuable problems that can be solved
by the ranks and inertias of matrices were neglected

c) Theorem 2.1(a)–(f) demonstrates such a simple fact that once certain expansion formulas for the partial inertia of a Hermitian matrix
expression are derived, we can use them to explicitly characterize
the definiteness of the matrix expression. In Sections 3 and 4, we
shall use Theorem 2.1(a)–(e) to obtain various conditional inequalities for orthogonal projectors and their operations in the Löwner
partial ordering.

d) The essential part of employing (2.15) is to find two different Hermitian matrices N_1 and N_2 that are congruent to M. Theoretically
speaking, for any given matrices, we can use them to construct some
block Hermitian matrices and to establish certain Hermitian congruence transformations for the block matrices, as demonstrated in
(2.16)–(2.22). These Hermitian congruence transformations may or
may not produce some informative reduced forms. Hence, they can
or cannot be used to derive some acceptable results on algebraic
properties of the given matrices. The three pairs of Hermitian congruence transformations in (2.17)–(2.22) are established according
to the given matrix polynomials $A - A^2$, $I_m - A^2$ and $A - A^3$ and their
three factors, and the main features of the right-hand sides of the six
congruence transformations in (2.17)–(2.22) are block diagonal or
skew diagonal. So that we are able to obtain the equalities for the

inertias of $A - A^2$, $I_m - A^2$ and $A - A^3$ by using (1.8) and (1.9) to the six congruence transformations. This method was also successfully used in the author's recent paper [50].

e) In the investigations of Hermitian matrices and applications, a popular method is using the spectral decompositions of the Hermitian matrices and their eigenvalues. However, this method is not so efficient when some matrices and their operations occur in the problems considered. The derivations of (2.6)–(2.13) show that algebraic properties of Hermitian matrices can also be derived without using the spectral decomposition of A and its eigenvalues.

Because (2.6)–(2.13) have such nice forms, it is worth trying to extend the expansion formulas to some general settings. For instance, it can be derived from (1.7) that the matrix pencil $\lambda A - \lambda^2 A^2$ satisfies

$$i_\pm(\lambda A - \lambda^2 A^2) = \begin{cases} i_\pm(A - \lambda A^2) & \text{if } \lambda > 0, \\ i_\mp(A - \lambda A^2) & \text{if } \lambda < 0 \end{cases}$$

(2.23)

For any Hermitian matrix A. Hence, we obtain from (2.6) and (2.7) the following result on the partial inertia of $A - \lambda A^2$ and the corresponding conditional matrix inequalities.

Corollary 2.3: Let $A \in \mathbb{C}_H^m$ and λ be a real number. Then,

$$i_+(A - \lambda A^2) = i_+(I_m - \lambda A) + i_+(A) - m \quad \text{for } \lambda > 0,$$

(2.24)

$$i_-(A - \lambda A^2) = i_-(I_m - \lambda A) + i_-(A) \quad \text{for } \lambda > 0,$$

(2.25)

$$i_+(A - \lambda A^2) = i_-(I_m - \lambda A) + i_+(A) \quad \text{for } \lambda < 0,$$

(2.26)

$$i_-(A - \lambda A^2) = i_+(I_m - \lambda A) + i_-(A) - m \quad \text{for } \lambda < 0.$$

(2.27)

Hence, under $\lambda > 0$

a) $A - \lambda A^2 > 0$ ($A - \lambda A^2 \geq 0$) if and only if $I_m > \lambda A > 0$ ($I_m \geq \lambda A \geq 0$).

b) $A - \lambda A^2 < 0$ ($A - \lambda A^2 \leq 0$) if and only if $i_-(I_m - \lambda A) + i_-(A) = m$ ($i_+(I_m - \lambda A) + i_+(A) = m$).

c) Under $\lambda < 0$,

d) $A - \lambda A^2 > 0$ $(A - \lambda A^2 \geq 0)$ if and only if $i_-\ (I_m - \lambda A) + i_+\ (A) = m$ $(i_+\ (I_m - \lambda A) + i_-\ (A) = m)$.

e) $A - \lambda A_2 < 0$ $(A - \lambda A_2 \leq 0)$ if and only if $I_m > \lambda A > 0$ $(I_m \geq \lambda A \geq 0)$.

Expansion formula for the partial inertia of a general quadratic matrix polynomial is given below.

Theorem 2.4: Let $A \in \mathbb{C}_H^m$, and let λ and μ be two real numbers with $\lambda\mu \neq 0$ and $\lambda < \mu$. Then,

$$i_+\big[(\lambda I_m - A)(\mu I_m - A)\big] = i_+(\lambda I_m - A) + i_-(\mu I_m - A), \quad (2.28)$$

$$i_-\big[(\lambda I_m - A)(\mu I_m - A)\big] = i_-(\lambda I_m - A) + i_+(\mu I_m - A) - m.$$

$$(2.29)$$

Hence,

a) $(\lambda I_m - A)\ (\mu I_m - A) > 0$ if and only if $i_+\ (\lambda I_m - A) + i_-\ (\mu I_m - A) = m$.

b) $(\lambda I_m - A)\ (\mu I_m - A) \geq 0$ if and only if $i_-\ (\lambda I_m - A) + i_+\ (\mu I_m - A) = m$.

c) $(\lambda I_m - A)\ (\mu I_m - A) < 0$ if and only if $\lambda Im < A < \mu I_m$.

d) $(\lambda I_m - A)\ (\mu I_m - A) \leq 0$ if and only if $\lambda I_m \leq A \leq \mu I_m$.

Proof: Let $B = \lambda I_m - A$ and $t = \mu - \lambda > 0$. Then, $(\lambda I_m - A)\ (\mu I_m - A)$ can be written as

$$(\lambda I_m - A)(\mu I_m - A) = B(tI_m + B) = tB + B^2. \quad (2.30)$$

Applying (2.24) and (2.25) to (2.30) gives

$$i_+\big(tB + B^2\big) = i_-\big[(-B) - t^{-1}(-B)^2\big] = i_-\big(I_m + t^{-1}B\big) + i_+(B)$$
$$= i_+(\lambda I_m - A) + i_-(\mu I_m - A),$$
$$i_-\big(tB + B^2\big) = i_+\big[(-B) - t^{-1}(-B)^2\big] = i_+\big(I_m + t^{-1}B\big) + i_-(B) - m$$
$$= i_-(\lambda I_m - A) + i_+(\mu I_m - A) - m,$$

As required for (2.28) and (2.29). Results (a)–(d) follow from (2.28), (2.29) and Lemma 1.1.

Eqs. (2.28) and (2.29) can be combined as

$$i_{\pm}\left[(\lambda I_m - A)(\mu I_m - A)\right] = i_{\pm}(\lambda I_m - A) + i_{\mp}(\mu I_m - A) - i_{\pm}\left[(\lambda - \mu)I_m\right]$$

(2.31)

for any $\lambda\mu \neq 0$ and $\lambda < \mu$.

Also note that $A - A^3$ is a special case of the matrix polynomial

$$f(A) = (\lambda_1 I_m - A)(\lambda_2 I_m - A)(\lambda_3 I_m - A).$$

Hence, it is necessary to give some expansion formulas for the partial inertia of f(A) in terms of the partial inertias of $\lambda_1 I_m - A$, $\lambda_2 I_m - A$ and λ_3 Im − A when A is Hermitian and $\lambda_1 < \lambda_2 < \lambda_3$. Further, recall that any real polynomial f (x) has certain irreducible factorizations. Thus, it would be of interest to establish some expansion formulas for the partial inertia of a matrix polynomial f (A) through its irreducible factorizations when A is Hermitian.

EXPANSION FORMULAS FOR RANKS/INERTIAS OF MATRIX PENCILS GENERATED FROM TWO ORTHOGONAL PROJECTORS

Idempotent matrices and its special class—Hermitian idempotent matrices (orthogonal projectors) are considered as a simple but important class of matrices. In any case, idempotent matrices or orthogonal projectors and their algebraic aspects are interesting in themselves. Linear combinations of idempotent matrices or orthogonal projectors and their applications, as well as polynomials in idempotent matrices or orthogonal projectors and their algebraic properties were widely considered in the literature; see, e.g., [3–8,10,11,17,21,34,40,43,44,52,53,55,57,56,61]. It is well known from the spectral decomposition of a Hermitian matrix that any Hermitian matrix can be decomposed as a linear combination of certain mutually disjoint orthogonal projectors, or a linear combination of at most four orthogonal projections; see [39]. These facts prompt us to consider ranks/inertias of linear combinations of two or more orthogonal projectors. Just as the Hermitian congruence transformations given in (2.17)–(2.22), we are also able to construct some Hermitian congruence transformations for block matrices consisting of orthogonal projectors and their operations, and then use them to derive some expansion formulas for the rank/inertia of the linear combination of two orthogonal projectors of the same size.

For any two given matrices A and B of the same size and any two scalars a and b, the linear combination aA + bB is often called a matrix pencil. The theory of matrix pencils is widely used in contemporary linear algebra and its applications; see, e.g., [2, 15, 19, 20, 31, 33, 46, 47, 58–60]. From the algebraic point of view, the mechanism of a general matrix pencil is not easy to distinguish. If, however, both P and Q are a pair of idempotent matrices or orthogonal projectors of the same size, then the matrix pencil aP + bQ has many nice algebraic properties. By making use of certain elementary matrix operations for block matrices, it was shown in [55, 57] that for any pair of idempotent matrices P and Q of the same size, and two scalars a and b such that $ab \neq 0$ and $a + b \neq 0$, the following expansion formulas for the ranks of aP + bQ, P − Q, aPQ + bQ P and P Q − Q P hold.

$$r(aP + bQ) = r\begin{bmatrix} P & Q \\ Q & 0 \end{bmatrix} - r(Q),$$

(3.1)

$$r(P - Q) = r\begin{bmatrix} P \\ Q \end{bmatrix} + r[P, Q] - r(P) - r(Q),$$

(3.2)

$$r(aPQ + bQ P) = r(P + Q) + r(PQ) + r(Q P) - r(P) - r(Q),$$

(3.3)

$$r(PQ - Q P) = r(P - Q) + r(I_m - P - Q) - m,$$

(3.4)

$$r(PQ - Q P) = r\begin{bmatrix} P \\ Q \end{bmatrix} + r[P, Q] + r(PQ) + r(Q P) - 2r(P) - 2r(Q).$$

(3.5)

These rank formulas were derived from some block matrix equalities consisting of P and Q . For instance, (3.4) can be derived from the following two decompositions

$$
\begin{bmatrix} I_m & P+Q-I_m \\ P-Q & 0 \end{bmatrix} = \begin{bmatrix} I_m & I_m-2P \\ 0 & I_m \end{bmatrix} \begin{bmatrix} 0 & P+Q-I_m \\ P-Q & 0 \end{bmatrix} \begin{bmatrix} I_m & 0 \\ I_m-2Q & I_m \end{bmatrix},
$$

$$
\begin{bmatrix} I_m & P+Q-I_m \\ P-Q & 0 \end{bmatrix} = \begin{bmatrix} I_m & 0 \\ P-Q & I_m \end{bmatrix} \begin{bmatrix} I_m & 0 \\ 0 & QP-PQ \end{bmatrix} \begin{bmatrix} I_m & P+Q-I_m \\ 0 & I_m \end{bmatrix}.
$$

The rank expansion formulas in (3.1)–(3.5) can be used to characterize the nonsingularity of the matrix pencils aP + bQ and aPQ + bQ P, as well as to the equalities P = Q and P Q = Q P. In addition, for any pair of orthogonal projectors P and Q, the following rank equalities

$$
r(PQ-QP) = 2r(PQ-PQP) = 2r(PQ-QPQ) = 2r\left[PQ-(PQ)^2\right]
$$

$$
(3.6)
$$

Hold; see [14]. Eq. (3.6) obviously implies that

P Q is an orthogonal projector $\Leftrightarrow$

$$(P\,Q)^2 = PQ \Leftrightarrow P\,Q = Q\,P \Leftrightarrow P\,Q = PQP = Q\,P\,Q. \tag{3.7}$$

In the investigation of orthogonal projectors, much attention has been paid to (simultaneous) decompositions of projectors and their operations. The well-known CS decomposition asserts that for a pair of orthogonal projectors P and Q of order m, there exists a unitary matrix U such that

$$
P = U \operatorname{diag}\{I_{k_1}, 0_{k_2}, I_{k_3}, I_{k_4}, 0_{k_5}, 0_{k_6}\}U^*, \qquad Q = U \operatorname{diag}\left\{ \begin{bmatrix} C^2 & CS \\ SC & S^2 \end{bmatrix}, I_{k_3}, 0_{k_4}, I_{k_5}, 0_{k_6} \right\} U^*,
$$

$$
(3.8)
$$

Where C and S are two positive diagonal matrices such that

$$
C^2 + S^2 = I_{k_1}, \qquad k_1 + k_3 + k_4 = r(P), \qquad r\begin{bmatrix} C^2 & CS \\ SC & S^2 \end{bmatrix} + k_3 + k_5 = r(Q), \qquad k_1 + \cdots + k_6 = m;
$$

see, e.g., [10,16,24]. Hence, the two products P Q and Q P can be represented as

$$P Q = U \operatorname{diag} \left\{ \begin{bmatrix} C^2 & CS \\ 0 & 0 \end{bmatrix}, I_{k_3}, 0 \right\} U^*, \qquad Q P = U \operatorname{diag} \left\{ \begin{bmatrix} C^2 & 0 \\ SC & 0 \end{bmatrix}, I_{k_3}, 0 \right\} U^*,$$

(3.9)

Which was shown in [22]. Further

$$P + Q = U \operatorname{diag} \left\{ \begin{bmatrix} I_{k_1} + C^2 & CS \\ SC & S^2 \end{bmatrix}, 2I_{k_3}, I_{k_4}, I_{k_5}, 0_{k_6} \right\} U^*,$$

(3.10)

$$P - Q = U \operatorname{diag} \left\{ \begin{bmatrix} S^2 & -CS \\ -SC & -S^2 \end{bmatrix}, 0_{k_3}, I_{k_4}, -I_{k_5}, 0_{k_6} \right\} U^*.$$

(3.11)

These decompositions can be used to obtain various algebraic properties for a pair of orthogonal projectors and their operations. The processes are, however, somewhat complicated in most situations. For three or more orthogonal projectors of the same size, it is hard to establish some simultaneous decomposition with informative structures. In this situation, we can only use the conventional operations for matrices to derive equalities for ranks/inertias of orthogonal projectors and their operations.

Note that the matrix pencil aP + bQ is Hermitian if P and Q are both orthogonal projectors and a and b are both real. In this case, the rank/inertia of aP + bQ may vary with respect to choice of the two real scalars a and b. Motivated by (3.1) and (3.2), we obtain the following results on the rank/inertia of aP + bQ and their consequences.

Theorem 3.1: Let P, Q $\in \mathbb{C}_{OP}^m$, a and b be two real numbers with ab $\neq$ 0 and a + b $\neq$ 0. Then

$$i_{\pm}(aP + bQ) = i_{\pm} \begin{bmatrix} tP & Q \\ Q & 0 \end{bmatrix} + i_{\mp}(tP) - i_{\mp}(aP) - i_{\mp}(bQ),$$

(3.12)

$$i_{\pm}(aP + bQ) = i_{\pm} \begin{bmatrix} tQ & P \\ P & 0 \end{bmatrix} + i_{\mp}(tQ) - i_{\mp}(bQ) - i_{\mp}(aP),$$

(3.13)

$$r(aP + bQ) = r(P + Q) = r[P, Q],$$ (3.14)

Where $t = a^{-1} + b^{-1}$. In particular,

a) $i_+(aP + bQ) = r[P, Q]$ and $i_-(aP + bQ) = 0$ if $a > 0$ and $b > 0$.

b) $i_+(aP + bQ) = 0$ and $i_-(aP + bQ) = r[P, Q]$ if $a < 0$ and $b < 0$.

c) $i_+(aP + bQ) = r(P)$ and $i_-(aP + bQ) = r[P, Q] - r(P)$ if $a > 0$, $b < 0$ and $a + b > 0$. In this case, $aP + bQ \leq 0$ if and only if $P = 0$; $aP + bQ \geq 0$ if and only if $\Re(Q) \subseteq \Re(P)$.

d) $i_+(aP + bQ) = r[P, Q] - r(Q)$ and $i_-(aP + bQ) = r(Q)$ if $a > 0$, $b < 0$ and $a + b < 0$. In this case, $aP + bQ \leq 0$ if and only if $\Re(P) \subseteq \Re(Q)$; $aP + bQ \geq 0$ if and only if $Q = 0$.

e) $i_+(aP + bQ) = r(Q)$ and $i_-(aP + bQ) = r[P, Q] - r(Q)$ if $a < 0$, $b > 0$ and $a + b > 0$. In this case, $aP + bQ \leq 0$ if and only if $Q = 0$; $aP + bQ \geq 0$ if and only if $\Re(P) \subseteq \Re(Q)$.

f) $i+(aP + bQ) = r[P, Q] - r(P)$ and $i-(aP + bQ) = r(P)$ if $a < 0$, $b > 0$ and $a + b < 0$. In this case, $aP + bQ \leq 0$ if and only if $\Re(Q) \subseteq \Re(P)$; $aP + bQ \geq 0$ if and only if $P = 0$.

g) The pencil $aP + bQ$ is non-singular $\Leftrightarrow P + Q$ is non-singular $\Leftrightarrow r[P, Q] = m$. Proof

Proof: Let

$$M = \begin{bmatrix} -aP & 0 & aP \\ 0 & -bQ & bQ \\ aP & bQ & 0 \end{bmatrix}, \quad U = \begin{bmatrix} I_m & 0 & 0 \\ 0 & I_m & 0 \\ I_m & I_m & I_m \end{bmatrix}, \quad V = \begin{bmatrix} I_m & 0 & -\frac{a}{2b}P \\ 0 & I_m & 2^{-1}P \\ \frac{b}{a+b}I_m & 0 & I_m - \frac{a}{2(a+b)}P \end{bmatrix}.$$

Then, M is obviously Hermitian. It is also easy to verify that both U and V are nonsingular, and

$$UMU^* = \begin{bmatrix} -aP & 0 & 0 \\ 0 & -bQ & 0 \\ 0 & 0 & aP + bQ \end{bmatrix}, \quad VMV^* = \begin{bmatrix} -\frac{a}{b}(a+b)P & 0 & 0 \\ 0 & 0 & bQ \\ 0 & bQ & \frac{ab}{a+b}P \end{bmatrix}.$$

Applying (1.5)–(1.9) to the two equalities yields

$$i_\pm(M) = i_\pm(-aP) + i_\pm(-bQ) + i_\pm(aP + bQ) = i_\pm\left(-\frac{a}{b}(a+b)P\right) + i_\pm\begin{bmatrix} 0 & bQ \\ bQ & \frac{a}{b}(a+b)P \end{bmatrix},$$

$$i_\pm(aP + bQ) = i_\pm \begin{bmatrix} (a^{-1} + b^{-1})P & Q \\ Q & 0 \end{bmatrix} + i_\mp [(a^{-1} + b^{-1})P] - i_\mp(aP) - i_\mp(bQ),$$

Where the two scalars $\dfrac{a}{b}(a + b)$ and $a^{-1} + b^{-1}$ have the same sign. Hence, (3.12) holds. Eq. (3.13) can be shown similarly. By (1.16) and (1.17),

$$i_+ \begin{bmatrix} tP & Q \\ Q & 0 \end{bmatrix} = r[P, Q], \qquad i_- \begin{bmatrix} tQ & P \\ P & 0 \end{bmatrix} = r(P) \quad \text{for } t > 0,$$

$$(3.15)$$

$$i_+ \begin{bmatrix} tP & Q \\ Q & 0 \end{bmatrix} = r(Q), \qquad i_- \begin{bmatrix} tQ & P \\ P & 0 \end{bmatrix} = r[P, Q] \quad \text{for } t < 0,$$

$$(3.16)$$

$$r \begin{bmatrix} tP & Q \\ Q & 0 \end{bmatrix} = r[P, Q] + r(Q) \quad \text{for } t \neq 0.$$

$$(3.17)$$

Adding the two equalities in (3.12) and applying (3.17) leads to (3.14). Applying Lemma 1.1, (1.7), (3.15) and (3.16) to (3.12) or (3.13) leads to (a)–(g).

By recalling that $r(A) = \text{tr}(A)$ if A is idempotent, we can replace the $r(P)$ and $r(Q)$ in Theorem 3.1 with $\text{tr}(P)$ and $\text{tr}(Q)$, respectively. Also, by recalling that $r(A) = r(B)$ and $\mathcal{R}(A) \subseteq \mathcal{R}(B)$ if and only if $\mathcal{R}(A) = \mathcal{R}(B)$, we can rewrite the rank formula in (3.14) as

$$\mathcal{R}(aP + bQ) = \mathcal{R}[P, Q].$$

A special case of the pencil $aP + bQ$ is the difference $P - Q$, for which we have the following result (see also [50, Corollary 3.16]).

Theorem 3.2: Let $P, Q \in \mathbb{C}^m_{\text{OP}}$. Then,

$$i_+(P - Q) = r[P, Q] - r(Q) = r(P - PQ),$$

$$(3.18)$$

$$i_-(P - Q) = r[P, Q] - r(P) = r(Q - PQ),$$

$$(3.19)$$

$$r(P - Q) = 2r[P, Q] - r(P) - r(Q) = r(P - PQ) + r(PQ - Q),$$

$$(3.20)$$

$s(P - Q) = r(P) - r(Q)$. $\qquad$ (3.21)

Hence,

a) $P - Q$ is nonsingular if and only if $r[P, Q] = r(P) + r(Q) = m$.

b) $P = Q$ if and only if $\Re (P) = \quad (Q)$.

c) $P > Q$ ($P < Q$) if and only if $P = I_m$ and $Q = 0$ ($P = 0$ and $Q = I_m$).

d) $P \geq Q$ ($P \leq Q$) if and only if $\Re (Q) \subseteq \Re (P)$ ($\Re (P) \subseteq \Re (Q)$).

e) $r(P - Q) = r(P) + r(Q) \Leftrightarrow i_+(P - Q) = r(P) \Leftrightarrow i_-(P - Q) = r(Q) \Leftrightarrow$
   $\Re (P) \cap \Re (Q) = \{0\}$.

f) $r(P - Q) = r(P) - r(Q) \Leftrightarrow i_+(P - Q) = r(P) - r(Q) \Leftrightarrow i_-(P - Q) = 0$
   $\Leftrightarrow \Re (Q) \subseteq \Re (P)$.

g) The signature of $P - Q$ is zero if and only if $r(P) = r(Q)$.

The results in Theorems 3.1 and 3.2 can also be given in some alternative forms. For instance, if P is an orthogonal projector of order m, then the difference $I_m - P$ is both Hermitian and idempotent, and thus it is an orthogonal projector as well and is often called the complementary orthogonal projector of P. Applying Theorem 3.2 to the difference $(I_m - P) - Q$ when both P and Q are orthogonal projector of order m, we obtain the following expansion formulas and their consequences.

Corollary 3.3: Let $P, Q \in C_{OP}^m$. Then

$$i_+(I_m - P - Q) = m - r(P) - r(Q) + r(PQ), \qquad (3.22)$$

$$i_-(I_m - P - Q) = r(PQ), \qquad (3.23)$$

$$r(I_m - P - Q) = m - r(P) - r(Q) + 2r(PQ), \qquad (3.24)$$

$$s(I_m - P - Q) = m - r(P) - r(Q). \qquad (3.25)$$

Hence,

a) $P + Q$ has t eigenvalues equal to 1, where $t = r(P) + r(Q) - 2r(PQ)$.

b) $I_m - P - Q$ is nonsingular if and only if $r(PQ) = r(P) = r(Q)$.

c) $P + Q = I_m \Leftrightarrow r(P) + r(Q) = m$ and $PQ = 0$.

d) $I_m - P - Q > 0$ if and only if $P = Q = 0$.

e) $I_m - P - Q < 0$ if and only if $P = Q = I_m$.

f) $I_m - P - Q \geq 0$ if and only if $PQ = 0$.

g) $I_m - P - Q \leq 0$ if and only if $r(PQ) = r(P) + r(Q) - m$.

h) The signature of $I_m - P - Q$ is zero if and only if $r(P) + r(Q) = m$.

Proof: Applying (3.18) and (3.19) to the difference $(I_m - P) - Q$ yields

$$i_+(I_m - P - Q) = r[I_m - P, Q] - r(Q),$$
(3.26)

$$i_-(I_m - P - Q) = r[I_m - P, Q] - r(I_m - P) = r[I_m - P, Q] - m + r(P).$$
(3.27)

Applying (1.1) to $[I_m - P, Q]$ and simplifying, we obtain

$$r(I_m - P + Q) = r[I_m - P, Q] = r(I_m - P) + r[Q - (I_m - P)Q] = m - r(P) + r(PQ).$$
(3.28)

Inserting (3.28) into (3.26) and (3.27) produces (3.22) and (3.23). Eqs. (3.24) and (3.25) follow from (3.22) and (3.23). Results (a)–(h) follow from (3.22)–(3.25) and Lemma 1.1.

For any two elements a and b in a ring, the two expressions ab − ba and ab + ba are often called the commutator and anti-commutator of a and b, respectively. The commutator and anti-commutator of two elements and their algebraic properties have been an attractive topic in no commutative algebra. Note that $PQ + QP$ is Hermitian if both P and Q ate Hermitian, and that $PQ + QP$ can be written as $PQ + QP = (P + Q)^2 - (P + Q)$ if both P and Q are orthogonal projectors. Hence, we are able to derive from Theorems 2.1, 3.1 and Corollary 3.3 the following results

Theorem 3.4: Let $P, Q \in \mathbb{C}_{OP}^m$. Then

$$i_+[(P + Q)^2 - (P + Q)] = i_+(PQ + QP) = r(PQ),$$
(3.29)

$$i_-[(P+Q)^2 - (P+Q)] = i_-(PQ + QP) = r[P, Q] - r(P) - r(Q) + r(PQ),$$
(3.30)

$$r[(P+Q)^2 - (P+Q)] = r(PQ + QP) = r[P, Q] - r(P) - r(Q) + 2r(PQ),$$
(3.31)

$$s[(P+Q)^2 - (P+Q)] = s(PQ + QP) = r(P) + r(Q) - r[P, Q],$$
(3.32)

$$2i_-(PQ + QP) = r(PQ - QP).$$
(3.33)

Hence,

a) $(P + Q)^2 - (P + Q)$ is non-singular $\Leftrightarrow$ P Q + Q P is non-singular $\Leftrightarrow$ r[P, Q] = m and r(P Q) = r(P) = r(Q).

b) $(P + Q)^2 >$ P + Q $\Leftrightarrow$ P Q + Q P > 0 $\Leftrightarrow$ P = Q = I$_m$.

c) $(P + Q)^2 \geq$ P + Q $\Leftrightarrow$ P Q + Q P $\geq$ 0 $\Leftrightarrow$ P Q = Q P $\Leftrightarrow$ r[P, Q] = r(P) + r(Q) − r(P Q) $\Leftrightarrow$ P Q $\in \mathbb{C}^m_{OP}$.

d) $(P + Q)^2 \leq$ P + Q $\Leftrightarrow$ (P + Q) 2 = P + Q $\Leftrightarrow$ P Q + Q P = 0 $\Leftrightarrow$ P Q = 0.

e) The signature of P Q + Q P is zero if and only if $\mathfrak{R}$ (P) $\cap$ $\mathfrak{R}$ (Q) = {0}.

Proof: Note that $(P + Q)^2 - (P + Q) = P Q + Q P$. Consequently, applying (2.6) and (2.7) to this $(P + Q)^2 - (P + Q)$ gives

$$i_+(PQ + QP) = i_+[(P+Q)^2 - (P+Q)] = i_-(P+Q) + i_-(I_m - P - Q),$$
(3.33)

$$i_-(PQ + QP) = i_-[(P+Q)^2 - (P+Q)] = i_+(P+Q) + i_+(I_m - P - Q) - m.$$
(3.34)

Substituting Lemma 3.1(a), (3.22) and (3.23) into (3.34) and (3.35), we obtain (3.29) and (3.30). Eqs. (3.31) and (3.32) follow from (3.29) and (3.30). Comparing (3.5) and (3.30) yields (3.33). Results (a)–(e) follow from (3.7), (3.29)–(3.33), and Lemma 1.1.

The equivalence of P Q + Q P $\geq$ 0 and P Q $\in \mathbb{C}^m_{OP}$ in Theorem 3.4(c) was given in [23]. Eqs. (3.29) and (3.30) are two expansion formulas for calculating the partial inertia of the anti-commutator of two orthogonal projectors. These two formulas show that the anti-commutator of two orthogonal projectors may have positive and negative eigenvalues simultaneously if both P Q $\neq$ 0 and r [P, Q] > r (P) + r (Q) − r (P Q). Thus, P Q + Q P are not definite in this case. A challenging task on the anti-commutator of two Hermitian matrices A and B is to give the distribution of the inertia triplet of A B + B A under the conditions A $\geq$ 0 and B $\geq$ 0.

A generalization of (3.29) and (3.30) is given below.

Theorem 3.5: Let P, Q $\in \mathbb{C}^m_{OP}$. Then

$$i_{\pm}\left[(PQ)^k + (QP)^k\right] = i_{\pm}(PQ + QP)$$

(3.36)

For any integer k $\geq$ 2.

Proof: It is easy to derive by induction that

$$(PQ)^k + (QP)^k = (P+Q)(P+Q-I_m)^{2k-1} = (P+Q-I_m)^{2k-1}(P+Q).$$

(3.37)

$$i_{\pm}\left[(PQ)^k + (QP)^k\right] = i_{\pm}\left[(P+Q)(P+Q-I_m)^{2k-1}\right] = i_{\pm}\left[(P+Q)(P+Q-I_m)\right]$$
$$= i_{\pm}(PQ+QP),$$

As required for (3.36).

By a similar approach, we are also able to obtain the following expansion formulas for the partial inertias of some Hermitian polynomials in two orthogonal projectors.

Theorem 3.6: Let P, Q $\in \mathbb{C}^m_{OP}$. Then,

$$i_+\left[(P+Q)^2 - I_m\right] = r(PQ),$$

(3.38)

$$i_-\left[(P+Q)^2 - I_m\right] = m - r(P) - r(Q) + r(PQ),$$ (3.39)

$$r\left[(P+Q)^2 - I_m\right] = m - r(P) - r(Q) + 2r(PQ),$$ (3.40)

$$s\left[(P+Q)^2 - I_m\right] = r(P) + r(Q) - m.$$ (3.41)

Hence,

a) $(P+Q)^2 - I_m$ is non-singular if and only if $r(PQ) = r(P) = r(Q)$.
b) $(P+Q)^2 - I_m > 0$ if and only if $P = Q = I_m$.
c) $(P+Q)^2 - I_m \geq 0$ if and only if $r(PQ) = r(P) + r(Q) - m$.
d) $(P+Q)^2 - I_m < 0$ if and only if $P = Q = 0$.
e) $(P+Q)^2 - I_m \leq 0$ if and only if $PQ = 0$.
f) $(P+Q)^2 = I_m$ if and only if $PQ = 0$ and $r(P) + r(Q) = m$.
g) The signature of $(P+Q)2 - I_m$ is zero if and only if $r(P) + r(Q) = m$.

Proof: Note that $(P+Q)^2 - I_m$ is Hermitian. Applying (2.9), (2.10), (3.22) and (3.23) to $(P+Q)^2 - I_m$, and simplifying by Theorem 3.1(a), (3.22) and (3.23), we obtain

$$i_+\left[(P+Q)^2 - I_m\right] = i_-(I_m + P + Q) + i_-(I_m - P - Q) = r(PQ),$$
$$i_-\left[(P+Q)^2 - I_m\right] = i_+(I_m + P + Q) + i_+(I_m - P - Q) - m = m - r(P) - r(Q) + r(PQ),$$

Establishing (3.38) and (3.39). Eqs. (3.40) and (3.41) follow from (3.38) and (3.39). Results (a)–(g) follow from (3.38)–(3.41) and Lemma 1.1.

Theorem 3.7: Let $P, Q \in \mathbb{C}^m_{OP}$. Then,

$$i_+\left[(P+Q)^3 - (P+Q)\right] = r(PQ),$$ (3.42)

$$i_-\left[(P+Q)^3 - (P+Q)\right] = r[P, Q] - r(P) - r(Q) + r(PQ),$$ (3.43)

$$r\left[(P+Q)^3 - (P+Q)\right] = r[P, Q] - r(P) - r(Q) + 2r(PQ),$$ (3.44)

$$s\left[(P+Q)^3 - (P+Q)\right] = r(P) + r(Q) - r[P, Q],$$ (3.45)

$$2i_-\left[(P+Q)^3-(P+Q)\right]=r(PQ-QP). \tag{3.46}$$

Hence,

a) $(P+Q)^3-(P+Q)$ is non-singular if and only if $r[P,Q]=m$ and $r(PQ)=r(P)=r(Q)$.

b) $(P+Q)^3 > (P+Q)$ if and only if $P=Q=I_m$.

c) $(P+Q)^3 \geq P+Q \Leftrightarrow PQ=QP \Leftrightarrow r[P,Q]=r(P)+r(Q)-r(PQ)$ $\Leftrightarrow PQ \in \mathbb{C}_{OP}^m$.

d) $(P+Q)^3 \leq P+Q \Leftrightarrow (P+Q)3 = P+Q \Leftrightarrow PQ=0$. (e) The signature of $(P+Q)3-(P+Q)$ is zero if and only if $\mathfrak{R}(P) \cap \mathfrak{R}(Q)=\{0\}$.

Proof: Note that $(P+Q)3-(P+Q)$ is Hermitian. Applying (2.12) to this expression and simplifying by Theorem 3.1(a), (3.22) and (3.23), we obtain

$$i_+\left[(P+Q)^3-(P+Q)\right]=i_-(P+Q)+i_+(I_m+P+Q)+i_-(I_m-P-Q)-m=r(PQ),$$
$$i_-\left[(P+Q)^3-(P+Q)\right]=i_+(P+Q)+i_-(I_m+P+Q)+i_+(I_m-P-Q)-m$$
$$=r[P,Q]-r(P)-r(Q)+r(PQ).$$

Establishing (3.42) and (3.43). Eqs. (3.44) and (3.45) follow from (3.42) and (3.43). Comparing (3.5) and (3.43) yields (3.46). Results (a)–(e) follow from (3.7), (3.42)–(3.46) and Lemma 1.1.

Observe that the right-hands of (3.29), (3.38) and (3.42) are the same, while the right-hands of (3.30) and (3.43) are the same. This fact allows us to conjecture that

$$i_+\left[(P+Q)^k-(P+Q)\right]=r(PQ), \qquad i_-\left[(P+Q)^k-(P+Q)\right]=r[P,Q]-r(P)-r(Q)+r(PQ)$$

For any integer $k \geq 2$

Theorem 3.8: Let $P, Q \in \mathbb{C}_{OP}^m$. Then

$$i_+\left[(P-Q)^2-(P-Q)\right]=r[P,Q]-r(P), \tag{3.47}$$

$$i_-\left[(P-Q)^2-(P-Q)\right]=r[P,Q]-r(P)-r(Q)+r(PQ), \tag{3.48}$$

$$r[(P - Q)^2 - (P - Q)] = 2r[P, Q] - 2r(P) - r(Q) + r(PQ),$$

(3.49)

$$s[(P - Q)^2 - (P - Q)] = r(Q) - r(PQ),$$

(3.50)

$$2i_-[(P - Q)^2 - (P - Q)] = r(PQ - QP).$$

(3.51)

Hence,

a) $(P - Q)_2 - (P - Q)$ is non-singular if and only if $2r[P, Q] = 2r(P) + r(Q) - r(PQ) + m$.
b) $(P - Q)^2 - (P - Q) > 0$ if and only if $P = 0$ and $Q = I_m$.
c) $(P - Q)^2 - (P - Q) \geq 0 \Leftrightarrow PQ = QP \Leftrightarrow r[P, Q] = r(P) + r(Q) - r(PQ) \Leftrightarrow PQ \in \mathbb{C}_{OP}^m$.
d) $(P - Q)^2 - (P - Q) \leq 0 \Leftrightarrow (P - Q)2 = (P - Q) \Leftrightarrow PQ = Q \Leftrightarrow \mathfrak{R}(Q) \subseteq \mathfrak{R}(P)$.
e) The signature of $(P - Q)^2 - (P - Q)$ is zero if and only if $r(Q) = r(PQ)$.

Proof: Note that $(P - Q)2 - (P - Q)$ is Hermitian. Applying (2.6) and (2.7) to $(P - Q)2 - (P - Q)$ and simplifying by (3.18), (3.19) and (3.28), we obtain

$$i_+[(P - Q)^2 - (P - Q)] = i_-(P - Q) + i_-(I_m - P + Q) = r[P, Q] - r(P),$$
$$i_-[(P - Q)^2 - (P - Q)] = i_+(P - Q) + i_+(I_m - P + Q) - m$$
$$= r[P, Q] - r(Q) + m - r(P) + r(PQ) - m$$
$$= r[P, Q] - r(Q) - r(P) + r(PQ),$$

Establishing (3.47) and (3.48) Eqs (3.49) and (3.50) follow from (3.47) and (3.48). Comparing (3.5) and (3.48) yields (3.51). Results (a)–(e) follow from (3.7), (3.47)–(3.51) and Lemma 1.1.

Theorem 3.9: Let $P, Q \in \mathbb{C}_{OP}^m$. Then,

$$i_-[(P - Q)^2 - I_m] = r[(P - Q)^2 - I_m] = m - r(P) - r(Q) + 2r(PQ).$$

(3.52)

Hence,

a) $(P - Q)^2 - I_m$ is non-singular if and only if r (P Q) = r (P) = r (Q).

b) $(P - Q)^2 \geq I_m$ always holds.

c) $(P - Q)^2 \leq I_m \Leftrightarrow (P - Q) 2 = I_m \Leftrightarrow P Q = 0$ and r (P) + r (Q) = m.

d) $(P - Q)^2$ has t eigenvalues equal to 1, where t = r (P) + r (Q) − 2r (P Q).

Proof: Note that $(P - Q) 2 - I_m$ is Hermitian. Applying (2.9) and (2.10) to $(P - Q) 2 - I_m$ and simplifying by (3.18), (3.19) and (3.28), we obtain

$$i_+\left[(P - Q)^2 - I_m\right] = i_-(I_m + P - Q) + i_-(I_m - P + Q) = 0,$$
$$i_-\left[(P - Q)^2 - I_m\right] = i_+(I_m + P - Q) + i_+(I_m - P + Q) - m$$
$$= r[P, I_m - Q] + r[I_m - P, Q] - m = m - r(Q) - r(P) + 2r(PQ),$$

Theorem 3.10: Let $P, Q \in \mathbb{C}^m_{OP}$. Then,

$$i_\pm\left[(P - Q)^3 - (P - Q)\right] = i_\pm(QPQ - PQP) = r[P, Q] - r(P) - r(Q) + r(PQ),$$
$$(3.53)$$

$$r\left[(P - Q)^3 - (P - Q)\right] = r(QPQ - PQP) = 2r[P, Q] - 2r(P) - 2r(Q) + 2r(PQ),$$
$$(3.54)$$

$$s\left[(P - Q)^3 - (P - Q)\right] = s(QPQ - PQP) = 0,$$
$$(3.55)$$

$$r(QPQ - PQP) = r(PQ - QP).$$
$$(3.56)$$

Hence,

$$(P - Q)^3 \geq P - Q \quad \Leftrightarrow \quad (P - Q)^3 \leq P - Q \quad \Leftrightarrow \quad (P - Q)^3 = P - Q \quad \Leftrightarrow \quad QPQ = PQP$$
$$\Leftrightarrow \quad PQ = QP \quad \Leftrightarrow \quad r[P, Q] = r(P) + r(Q) - r(PQ) \quad \Leftrightarrow \quad PQ \in \mathbb{C}^m_{OP}.$$
$$(3.57)$$

Proof: Note that $(P - Q)^3 - (P - Q)$ is Hermitian and it is easily verified that $(P - Q)^3 - (P - Q) = QPQ - PQP$. Applying (2.12) to this equality and simplifying by (3.18), (3.19) and (3.28), we obtain

$$i_+\left[(P - Q)^3 - (P - Q)\right] = i_+(I_m + P - Q) + i_-(I_m - P + Q) + i_-(P - Q) - m$$
$$= r[P, Q] - r(P) - r(Q) + r(PQ),$$
$$i_-\left[(P - Q)^3 - (P - Q)\right] = i_-(I_m + P - Q) + i_+(I_m - P + Q) + i_+(P - Q) - m$$
$$= r[P, Q] - r(Q) - r(P) + r(PQ),$$

Theorem 3.11: Let $P, Q \in \mathbb{C}^m_{OP}$. Then,

$$i_+(QPQ + PQP) = r(QPQ + PQP) = r(PQ).\tag{3.58}$$

Proof: It is easy to verify that

$$PQP + QPQ = (P + Q)(P + Q - I_m)^2 = (P + Q - I_m)^2(P + Q).$$

Also note that $P + Q \geq 0$. Then, we find by (1.11) and (3.29) that

$$i_+(PQP + QPQ) = r(PQP + QPQ) = i_+\left[(P + Q)(P + Q - I_m)^2\right]$$
$$= i_+\left[(P + Q)(P + Q - I_m)\right] = i_+(PQ + QP) = r(PQ),$$

as required for (3.58).

Further, we have the following results on $P \pm PQP$ and $I_m - PQP$.

Theorem 3.12: Let $P, Q \in \mathbb{C}^m_{OP}$. Then

$$i_+(P + PQP) = r(P + PQP) = r(P),\tag{3.60}$$

$$i_+(P - PQP) = r(P - PQP) = r[P, Q] - r(Q),\tag{3.61}$$

$$i_+(Q + PQP) = r(Q + PQP) = r[P, Q] - r(P) + r(PQ),\tag{3.62}$$

$$i_+(Q - PQP) = r[P, Q] - r(P),\tag{3.63}$$

$$i_-(Q - PQP) = r[P, Q] - r(P) - r(Q) + r(PQ), \tag{3.64}$$

$$i_+(I_m - PQP) = r(I_m - PQP) = r[P, Q] - r(P) - r(Q) + m. \tag{3.65}$$

Hence,

a) $P \leq PQP \Leftrightarrow P = PQP \Leftrightarrow \mathfrak{R}(P) \subseteq \mathfrak{R}(Q)$.
b) $Q \leq PQP \Leftrightarrow Q = PQP \Leftrightarrow \mathfrak{R}(Q) \subseteq \mathfrak{R}(P)$.
c) $Q \geq PQP \Leftrightarrow r[P, Q] = r(P) + r(Q) - r(PQ)$.
d) PQP has t eigenvalues equal to 1, where $t = r(P) + r(Q) - r[P, Q]$.
e) $I_m \geq PQP$ always holds, and $I_m > PQP \Leftrightarrow \mathfrak{R}(P) \cap \mathfrak{R}(Q) = \{0\}$.

Proof: Note that $I_m + Q > 0$ and $I_m - Q \geq 0$. Therefore, $P(I_m \pm Q)P = P \pm PQP \geq 0$. Consequently, we obtain from (1.15) that

$$i_+(P + PQP) = r(P + PQP) = r(P),$$

$$i_+(P - PQP) = r(P - PQP) = r\begin{bmatrix} P & PQ \\ QP & Q \end{bmatrix} - r(Q)$$

$$= r\big([P, Q]^*[P, Q]\big) - r(Q) = r[P, Q] - r(Q),$$

As required for (3.60) and (3.61). Note that $Q + PQP \geq 0$. Therefore, applying (1.2), (1.4) and (3.14), and simplifying by EBMOs, we obtain

$$i_+(Q + PQP) = r(Q + PQP) = r[Q, PQP] = r[Q, PQ] = r[Q - PQ, PQ]$$

$$= r(Q - PQ) + r(PQ) = r[P, Q] - r(P) + r(PQ) \quad (\text{by (1.2)}),$$

As required for (3.62). Applying (1.14) and $Q - QPQPQ \geq 0$, and simplifying by EBMOs, we obtain

$$i_+(Q - PQP) = i_+\begin{bmatrix} Q & PQ \\ QP & Q \end{bmatrix} - i_+(Q) = i_+\begin{bmatrix} 0 & PQ - QPQ \\ QP - QPQ & Q - QPQPQ \end{bmatrix}$$

$$= r[QP - QPQ, Q - QPQPQ] \quad (\text{by (1.16)})$$

$$= r[QP - QPQ, Q - QPQ] = r[Q - QP, Q - QPQ]$$

$$= r[Q - QP, 0] = r[P, Q] - r(P) \quad (\text{by (1.2)}),$$

$$i_-(Q - PQP) = i_- \begin{bmatrix} Q & PQ \\ QP & Q \end{bmatrix} - i_-(Q) = i_- \begin{bmatrix} 0 & PQ - QPQ \\ QP - QPQ & Q - QPQPQ \end{bmatrix}$$

$$= r(PQ - QPQ) \quad (\text{by } (1.16))$$
$$= r[Q, PQ] - r(Q) \quad (\text{by } (1.2))$$
$$= r[Q - PQ, PQ] - r(Q) = r(Q - PQ) + r(PQ) - r(Q)$$
$$= r[P, Q] - r(P) - r(Q) + r(PQ) \quad (\text{by } (1.2)),$$

as required for (3.63) and (3.64). Applying (1.18) to $I_m - PQP$ and simplifying by (3.61), we obtain establishing (3.65).

$$i_+(I_m - PQP) = i_+ \begin{bmatrix} Q & QP \\ PQ & I_m \end{bmatrix} - i_+(Q) = i_+(Q - QPQ) + m - r(Q)$$

$$= r[P, Q] - r(P) - r(Q) + m,$$

$$i_-(I_m - PQP) = i_- \begin{bmatrix} Q & QP \\ PQ & I_m \end{bmatrix} - i_-(Q) = i_-(Q - QPQ) + I_-(I_m) = 0,$$

Hermitian matrix polynomials generated from two orthogonal projectors can be formulated arbitrarily. It seems from the previous results that some expansion formulas for the inertias of these Hermitian matrix polynomials can always be derived with some effort.

The matrix product $P\,Q$, the matrix pencil $aPQ + bQ\,P$ and the commutator $P\,Q - Q\,P$ are not necessarily Hermitian even both P and Q are orthogonal projectors. Hence, the rank formulas in (3.3)–(3.6) cannot be refined to the situations for inertia. However, if both P and Q are orthogonal projectors, the complex matrix $j\,(P\,Q - Q\,P)$ is Hermitian, where $j = \sqrt{-1}$. It can be seen from (3.5) that a reasonable conjecture on the partial inertia of $j\,(P\,Q - Q\,P)$ is given by

$$i_\pm[j(PQ - QP)] = r[P, Q] - r(P) - r(Q) + r(PQ).$$

EXPANSION FORMULAS FOR RANKS/INERTIAS OF MATRIX PENCILS GENERATED FROM THREE OR MORE ORTHOGONAL PROJECTORS

Matrix pencils can be generated from linear combinations of three or more matrices. In fact, the matrix expression $I_m - P - Q$ in Corollary 3.3 is a special case of the matrix pencils generated from the three orthogonal projectors I_m, P and Q. The mechanism of these pencils, however, is quite complicated in general. In what follows, we give some expansion formulas for the rank/inertia of a matrix pencil generated from three orthogonal projectors under some conditions.

Theorem 4.1: Assume that P, P1, P2 $\in \mathbb{C}^m_{OP}$ satisfy

$$\mathscr{R}(P_1) \subseteq \mathscr{R}(P) \quad and \quad \mathscr{R}(P_2) \subseteq \mathscr{R}(P), \tag{4.1}$$

And λ, $\lambda 1$ and $\lambda 2$ are nonzero real numbers. Also, denote

$$D = \text{diag}\{P_1, P_2\}, \qquad G = \begin{bmatrix} t_1 I_m & \lambda^{-1} I_m \\ \lambda^{-1} I_m & t_2 I_m \end{bmatrix}, \tag{4.2}$$

Where $t1 = \lambda^{-1}_1 + \lambda^{-1}$ and $t_2 = \lambda^{-1}_2 + \lambda^{-1}$. Then,

$$i_{\pm}(\lambda P + \lambda_1 P_1 + \lambda_2 P_2) = i_{\pm}(\lambda P) - i_{\mp}(\lambda_1 P_1) - i_{\mp}(\lambda_2 P_2) + i_{\mp}(DGD), \tag{4.3}$$

$$r(\lambda P + \lambda_1 P_1 + \lambda_2 P_2) = r(P) - r(P_1) - r(P_2) + r(DGD). \tag{4.4}$$

Hence,

a) If $G > 0$, i.e., $\lambda\lambda_1 (\lambda + \lambda_1) > 0$, $\lambda\lambda_2 (\lambda + \lambda_2) > 0$ and $\lambda\lambda_1\lambda_2 (\lambda + \lambda_1 + \lambda_2) > 0$, then

$$i_+(\lambda P + \lambda_1 P_1 + \lambda_2 P_2) = i_+(\lambda P) - i_-(\lambda_1 P_1) - i_-(\lambda_2 P_2), \tag{4.5}$$

$$i_-(\lambda P + \lambda_1 P_1 + \lambda_2 P_2) = i_-(\lambda P) - i_+(\lambda_1 P_1) - i_+(\lambda_2 P_2) + r(P_1) + r(P_2), \tag{4.6}$$

$$r(\lambda P + \lambda_1 P_1 + \lambda_2 P_2) = r(P). \tag{4.7}$$

b) If $G < 0$, i.e., $\lambda\lambda_1 (\lambda + \lambda_1) < 0$, $\lambda\lambda_2 (\lambda + \lambda_2) < 0$ and $\lambda\lambda_1\lambda_2 (\lambda + \lambda_1 + \lambda_2) > 0$, then

$$i_+(\lambda P + \lambda_1 P_1 + \lambda_2 P_2) = i_+(\lambda P) - i_-(\lambda_1 P_1) - i_-(\lambda_2 P_2) + r(P_1) + r(P_2),$$

$$(4.8)$$

$$i_-(\lambda P + \lambda_1 P_1 + \lambda_2 P_2) = i_-(\lambda P) - i_+(\lambda_1 P_1) - i_+(\lambda_2 P_2),\qquad (4.9)$$

$$r(\lambda P + \lambda_1 P_1 + \lambda_2 P_2) = r(P). \qquad (4.10)$$

c. $P - P_1 - P_2$ satisfies the following equalities

$$i_+(P - P_1 - P_2) = r(P) - r(P_1) - r(P_2) + r(P_1 P_2), \qquad (4.11)$$

$$i_-(P - P_1 - P_2) = r(P_1 P_2), \qquad (4.12)$$

$$r(P - P_1 - P_2) = r(P) - r(P_1) - r(P_2) + 2r(P_1 P_2), \qquad (4.13)$$

$$s(P - P_1 - P_2) = r(P) - r(P_1) - r(P_2). \qquad (4.14)$$

Hence,

i. $P - P_1 - P_2$ is non-singular if and only if $r(P) = r(P_1) + r(P_2) - 2r(P_1 P_2) + m$.

ii. $P - P_1 - P_2 \geq 0$ if and only if $P_1 P_2 = 0$.

iii. $P - P_1 - P_2 \leq 0$ if and only if $r(P) = r(P_1) + r(P_2) - r(P_1 P_2)$.

iv. $P = P_1 + P_2$ if and only if $P_1 P_2 = 0$ and $r(P) = r(P_1) + r(P_2)$.

v. The signature of $P - P_1 - P_2$ is zero if and only if $r(P) = r(P_1) + r(P_2)$.

d) $2P - P_1 - P_2$ satisfies the following equalities

$$i_+(2P - P_1 - P_2) = r(2P - P_1 - P_2) = r(P) - r(P_1) - r(P_2) + r[P_1, P_2].$$

$$(4.15)$$

Hence, $2P = P_1 + P_2$ if and only if $r(P) = r(P_1) + r(P_2) - r[P_1, P_2]$.

Proof: Let

$$
M = \begin{bmatrix} -\lambda^{-1}P & 0 & 0 & P \\ 0 & -\lambda_1^{-1}P_1 & 0 & P_1 \\ 0 & 0 & -\lambda_2^{-1}P_2 & P_2 \\ P & P_1 & P_2 & 0 \end{bmatrix},
$$

$$(4.16)$$

$$
U = \begin{bmatrix} I_m & 0 & 0 & 0 \\ 0 & I_m & 0 & 0 \\ 0 & 0 & I_m & 0 \\ \lambda I_m & \lambda_1 I_m & \lambda_2 I_m & I_m \end{bmatrix}, \qquad V = \begin{bmatrix} I_m & 0 & 0 & \frac{\lambda^{-1}}{2}P \\ -P_1 & I_m & 0 & -\lambda^{-1}P_1 \\ -P_2 & 0 & I_m & -\lambda^{-1}P_2 \\ 0 & 0 & 0 & I_m \end{bmatrix}.
$$

$$(4.17)$$

Then, it is easily verified that U and V are two non-singular matrices, and from $P P_1 = P_1 P = P_1$ and $P P_2 = P_2 P = P_2$ that

$$
UMU^* = \begin{bmatrix} -\lambda^{-1}P & 0 & 0 & 0 \\ 0 & -\lambda_1^{-1}P_1 & 0 & 0 \\ 0 & 0 & -\lambda_2^{-1}P_2 & 0 \\ 0 & 0 & 0 & \lambda P + \lambda_1 P_1 + \lambda_1 P_2 \end{bmatrix},
$$

$$(4.18)$$

$$
VMV^* = \begin{bmatrix} 0 & 0 & 0 & P \\ 0 & -(\lambda_1^{-1} + \lambda^{-1})P_1 & -\lambda^{-1}P_1 P_2 & 0 \\ 0 & -\lambda^{-1}P_2 P_1 & -(\lambda_2^{-1} + \lambda^{-1})P_2 & 0 \\ P & 0 & 0 & 0 \end{bmatrix}.
$$

$$(4.19)$$

Applying (1.5)–(1.9) to (4.18) and (4.19) yields

$$
i_{\pm}(M) = i_{\mp}(\lambda^{-1}P) + i_{\mp}(\lambda_1^{-1}P_1) + i_{\mp}(\lambda_2^{-1}P_2) + i_{\pm}(\lambda P + \lambda_1 P_1 + \lambda_2 P_2)
$$

$$
= r(P) + i_{\mp}\begin{bmatrix} t_1 P_1 & \lambda^{-1}P_1 P_2 \\ \lambda^{-1}P_2 P_1 & t_2 P_2 \end{bmatrix} = r(P) + i_{\mp}(DGD),
$$

Establishing (4.3) and (4.4).

If $G > 0$, then $DGD \geq 0$ and $i_+(DGD) = r(DGD) = r(D) = r(P_1) + r(P_2)$; if $G < 0$, then $DGD \leq 0$ and $i-(DGD) = r(DGD) = r(D) = r(P_1) + r(P_2)$. Hence, (4.3) and (4.4) reduce to (4.5)–(4.7) and (4.8)–(4.10), respectively. Let $\lambda = -\lambda_1 = -\lambda_2 = 1$. Then, it follows from (1.9) that

$$i_{\mp}(DGD) = i_{\mp} \begin{bmatrix} t_1 P_1 & \lambda^{-1} P_1 P_2 \\ \lambda^{-1} P_2 P_1 & t_2 P_2 \end{bmatrix} = i_{\pm} \begin{bmatrix} 0 & P_1 P_2 \\ P_2 P_1 & 0 \end{bmatrix} = r(P_1 P_2),$$

So that (4.3) reduces to (4.11) and (4.12). Eqs. (4.13) and (4.14) follow from (4.3) and (4.4). Results (i)–(IV) in (c) are direct consequences of (4.3)–(4.14). Let $2^{-1}\lambda = -\lambda_1 = -\lambda_2 = 1$. Then,

$$i_{\pm}(DGD) = i_{\pm} \begin{bmatrix} t_1 P_1 & \lambda^{-1} P_1 P_2 \\ \lambda^{-1} P_2 P_1 & t_2 P_2 \end{bmatrix} = i_{\pm} \begin{bmatrix} 2P_1 & 2P_1 P_2 \\ 2P_2 P_1 & 2P_2 \end{bmatrix} = i_{\pm}([P_1, P_2]^*[P_1, P_2]),$$

So that (4.3) reduces to (4.15).

The two range inclusions in (4.1) are a reasonable assumption on the relations among three orthogonal projectors, for instance, the orthogonal projectors onto the ranges of M = [A, B] and its two sub matrices A and B satisfy such conditions. The results in Theorem 4.1, as well as the constructions of (4.16)–(4.19) show that to obtain satisfactory expansion formulas for the inertia of a general matrix pencil generated from three or more orthogonal projectors is a challenging task. Setting $P = I_m$ in (4.3) and (4.4), we obtain the following result.

Corollary 4.2: Let P, Q $\in C_{OP}^m$, λ, λ_1 and λ_2 be nonzero real numbers, and denote

$$M = \begin{bmatrix} t_1 P & \lambda^{-1} PQ \\ \lambda^{-1} QP & t_2 Q \end{bmatrix}, \quad t_1 = \lambda_1^{-1} + \lambda^{-1} \text{ and } t_2 = \lambda_2^{-1} + \lambda^{-1}.$$

$$i_{\pm}(\lambda I_m + \lambda_1 P + \lambda_2 Q) = i_{\pm}(\lambda I_m) - i_{\mp}(\lambda_1 P) - i_{\mp}(\lambda_2 Q) + i_{\mp}(M), \qquad (4.20)$$

$$r(\lambda I_m + \lambda_1 P + \lambda_2 Q) = m - r(P) - r(Q) + r(M). \qquad (4.21)$$

Hence,

a) The linear matrix inequality $\lambda I_m + \lambda_1 P + \lambda_2 Q > 0$ (< 0) holds if and only if

$i_-(M) = m - i_+(\lambda I_m) + i_-(\lambda_1 P) + i_-(\lambda_2 Q)$ $(i_+(M) = m - i_-(\lambda I_m) + i_+(\lambda_1 P) + i_+(\lambda_2 Q))$.

a) The linear matrix inequality $\lambda I_m + \lambda_1 P + \lambda_2 Q \geq 0$ (≤ 0) holds if and only if

b) $i_+(M) = i_+(\lambda_1 P) + i_+(\lambda_2 Q) - i_-(\lambda I_m)$ $(i_-(M) = i_-(\lambda_1 P) + i_-(\lambda_2 Q) - i_+(\lambda I_m))$.

c) $\lambda I_m + \lambda_1 P + \lambda_2 Q = 0$ if and only if $r(M) = r(P) + r(Q) - m$.

d) $r(\lambda I_m + \lambda_1 P + \lambda_2 Q) < m$, i.e., $|\lambda I_m + \lambda_1 P + \lambda_2 Q| = 0$ if and only if $r(M) < r(P) + r(Q)$.

Applying Theorem 2.1 and Corollary 4.2 to some matrix polynomials of the matrix pencil $\lambda I_m + \lambda_1 P + \lambda_2 Q$, such as,

$$(\lambda I_m + \lambda_1 P + \lambda_2 Q) - (\lambda I_m + \lambda_1 P + \lambda_2 Q)^2, \qquad I_m - (\lambda I_m + \lambda_1 P + \lambda_2 Q)^2,$$

$$(\lambda I_m + \lambda_1 P + \lambda_2 Q) - (\lambda I_m + \lambda_1 P + \lambda_2 Q)^3$$

Will yield a variety of expansion formulas for the partial inertias of the matrix expressions. We leave this work for the reader.

When considering a pair of orthogonal projectors P, $Q \in \mathbb{C}_{OP}^m$, it is usually assumed that P and Q satisfy certain equalities, such as, P Q = Q P, PQP = P, PQP = Q, PQP = QPQ, etc. In these cases, all the linear combinations of the two orthogonal projectors and their possible products generate a finite-dimensional (non-)commutative algebra over the real or complex field. For instance, if the pair of orthogonal projectors satisfy P Q = Q P, then the corresponding matrix pencil

$$M = a_0 I_m + a_1 P + a_2 Q + a_3 P Q, \quad a_0, a_1, a_2, a_3 \in \mathbb{R} \tag{4.22}$$

Is Hermitian, and all these matrix pencils generate a commutative algebra up to four dimensions over the real number field $\mathbb{R}$ under the conventional addition and multiplication of matrices. This matrix algebra has many interesting properties. A remarkable universal similarity factorization equalityJ (USFE) associated with the pencil is given by

$$L \operatorname{diag}(M_1, M_2, M_3, M_4) L^{-1} = \operatorname{diag}(t_1 I_m, t_2 I_m, t_3 I_m, t_4 I_m), \tag{4.23}$$

Where

$$M_1 = a_0 I_m + a_1 P + a_2 Q + a_3 PQ,$$
$$M_2 = (a_0 + a_2)I_m + (a_1 + a_3)P - a_2 Q - a_3 PQ,$$
$$M_3 = (a_0 + a_1)I_m - a_1 P + (a_2 + a_3)Q - a_3 PQ,$$
$$M_4 = (a_0 + a_1 + a_2 + a_3)I_m - (a_1 + a_3)P - (a_2 + a_3)Q + a_3 PQ,$$
$$t_1 = a_0, \qquad t_2 = a_0 + a_2, \qquad t_3 = a_0 + a_1, \qquad t_4 = a_0 + a_1 + a_2 + a_3,$$

And

$$L = L^* = L^{-1} = \begin{bmatrix} L_1 & L_2 & L_3 & L_4 \\ L_2 & L_1 & -L_4 & -L_3 \\ L_3 & -L_4 & L_1 & -L_2 \\ L_4 & -L_3 & -L_2 & L_1 \end{bmatrix},$$

In which the four matrices

$$L_1 = I_m - P - Q + PQ, \qquad L_2 = Q - PQ, \qquad L_3 = P - PQ, \qquad L_4 = PQ$$

$$(4.24)$$

Satisfy

$$L_1 + L_2 + L_3 + L_4 = I_m, \qquad L_i^2 = L_i = L_i^*, \qquad L_i L_j = 0, \quad i \neq j, \ i, j = 1, \dots, 4,$$

$$(4.25)$$

$$r(L_1) = m - r(P) - r(Q) + r(PQ), \qquad r(L_2) = r(Q) - r(PQ), \qquad r(L_3) = r(P) - r(PQ);$$

$$(4.26)$$

See [53, 57,56]. It can be derived from (4.23) that the matrix pencil in (4.22) can be decomposed as the following linear combination of the four orthogonal projectors $L_1, \dots, L_4$:

$$M = t_1 L_1 + t_2 L_2 + t_3 L_3 + t_4 L_4 = \widehat{L} \operatorname{diag}(t_1 I_m, t_2 I_m, t_3 I_m, t_4 I_m)\widehat{L}^*, \qquad (4.27)$$

Where the row block matrix $L = [L_1, L_2, L_3, L_4]$ satisfies $LL* = I_m$. Notice the matrices $L_1 \dots L_4$ in (4.24) are four mutually disjoint orthogonal projectors, and the four scalars $t_1,\ldots,t_4$ are the eigenvalues of M. Therefore, (4.27) is in fact a closed-form spectral decomposition of the matrix pencil (4.22), which can also be called a disjoint orthogonal projection decomposition (DOPD) of the matrix pencil M in (4.22). Many consequences can be derived from the DOPD in (4.27). For instance,

i. The expansion formulas for the partial inertia of the matrix pencil M in (4.22) are

$$i_\pm(M) = i_\pm(t_1)r(L_1) + i_\pm(t_2)r(L_1) + i_\pm(t_3)r(L_3) + i_\pm(t_4)r(L_4)$$
$$= i_\pm(t_1)m + \left[i_\pm(t_3) - i_\pm(t_1)\right]r(P) + \left[i_\pm(t_2) - i_\pm(t_1)\right]r(Q)$$
$$+ \left[i_\pm(t_1) - i_\pm(t_2) - i_\pm(t_3) + i_\pm(t_4)\right]r(PQ).$$
(4.28)

ii. The expansion formula for the power of M in (4.22) can be decomposed as

$$M^k = t_1^k L_1 + t_2^k L_2 + t_3^k L_3 + t_4^k L_4$$
$$= t_1^k I_m + (t_3^k - t_1^k)P + (t_2^k - t_1^k)Q + (t_1^k - t_2^k - t_3^k + t_4^k)PQ$$
(4.29)

iii. For any integer $k \geq 2$
  a) The expansion formula for the exponential of M in (4.22) can be decomposed as

$$e^M = e^{t_1} L_1 + e^{t_2} L_2 + e^{t_3} L_3 + e^{t_4} L_4$$
$$= e^{t_1} I_m + (e^{t_3} - e^{t_1})P + (e^{t_2} - e^{t_1})Q + (e^{t_1} - e^{t_2} - e^{t_3} + e^{t_4})PQ.$$
(4.30)

iv. If $t_1 t_2 t_3 t_4 \neq 0$, then M in (4.22) is non-singular too, and the inverse of the M can be written as

$$M^{-1} = t_1^{-1} L_1 + t_2^{-1} L_2 + t_3^{-1} L_3 + t_4^{-1} L_4$$
$$= t_1^{-1} I_m + (t_3^{-1} - t_1^{-1})P + (t_2^{-1} - t_1^{-1})Q + (t_1^{-1} - t_2^{-1} - t_3^{-1} + t_4^{-1})PQ.$$
(4.31)

v. If $t_1 t_2 t_3 t_4 = 0$, then the Moore–Penrose inverse of M in (4.22) can be decomposed as

$$M^\dagger = t_1^\dagger L_1 + t_2^\dagger L_2 + t_3^\dagger L_3 + t_4^\dagger L_4$$
$$= t_1^\dagger I_m + (t_3^\dagger - t_1^\dagger)P + (t_2^\dagger - t_1^\dagger)Q + (t_1^\dagger - t_2^\dagger - t_3^\dagger + t_4^\dagger)PQ.$$
(4.32)

In particular, the DOPD of the matrix pencil $aP + aQ$ under $PQ = QP$ is

$$aP + aQ = a(P - PQ) + b(Q - PQ) + (a+b)PQ. \tag{4.33}$$

Hence, a, b and $a + b$ are eigenvalues of $aP + aQ$, and the following expansions hold

$$i_\pm(aP + bQ) = i_\pm(a)r(P) + i_\pm(b)r(Q) + \left[i_\pm(a+b) - i_\pm(a) - i_\pm(b)\right]r(PQ), \tag{4.34}$$

$$(aP + bQ)^k = a^k P + b^k Q + \left[(a+b)^k - a^k - b^k\right]PQ, \tag{4.35}$$

$$e^{aP+bQ} = e^a P + e^b Q + (e^{a+b} - e^a - e^b)PQ, \tag{4.36}$$

$$(aP + bQ)^\dagger = a^\dagger P + b^\dagger Q + \left[(a+b)^\dagger - a^\dagger - b^\dagger\right]PQ. \tag{4.37}$$

In addition, general solutions to the idempotent, tripotent and involutory equations $M^2 = M$, $M^2 = I_m$ and $M^3 = M$ can also be derived from (4.25) and (4.27). Some previous and recent work on idempotency, tripotency and involution of linear combinations of idempotent matrices can be found, e.g., in [3, 4, 7, 40, 53].

Without much effort, the above results on a pair of commutative orthogonal projectors can be extended to triple or more mutually commutative orthogonal projectors. For instance, if a triple orthogonal projectors P_1, P_2, $P_3 \in \mathbb{C}^m_{\rm OP}$ satisfy $P_i P_j = P_j P_i$, $i = 1, 2, 3$, then the corresponding eight-term matrix pencil

$$M = a_0 I_m + a_1 P_1 + a_2 P_2 + a_3 P_3 + a_{12} P_1 P_2 + a_{13} P_1 P_3 + a_{23} P_2 P_3 + a_{123} P_1 P_2 P_3 \tag{4.38}$$

is Hermitian as well, where $a_0, a_1, a_2, a_3, a_{12}, a_{13}, a_{23}, a_{123}$ are real numbers, and all the matrix pencils generate a commutative algebra up to eight dimensions over the real number field under the conventional addition and multiplication of matrices. A USFE associated with the pencil and the corresponding DOPD were given in [53], which can be used to produce various expansion formulas for the inertia, rank, power, expo-nential, inverse and Moore–Penrose inverse of M in (4.38).

Finally, we give an application of Theorem 3.2 to a k × k block Hermi-tian matrices consisting of orthogonal projectors.

Theorem 4.3: Let $P, P_1,..., P_k \in \mathbb{C}^m_{OP}$, and denote

$$
M = \begin{bmatrix} P - kP_1 & P & \cdots & P \\ P & P - kP_2 & \cdots & P \\ \vdots & \vdots & \ddots & \vdots \\ P & P & \cdots & P - kP_k \end{bmatrix}, \quad N = \begin{bmatrix} P & P_1 & 0 & \cdots & 0 \\ P & 0 & P_2 & \cdots & 0 \\ \vdots & \vdots & \vdots & \ddots & \vdots \\ P & 0 & 0 & \cdots & P_k \end{bmatrix}.
$$

$$(4.39)$$

Then,

$$
i_+(M) = r(N) - r(P_1) - \cdots - r(P_k),
$$

$$(4.40)$$

$$
i_-(M) = r(N) - r(P),
$$

$$(4.41)$$

$$
r(M) = 2r(N) - r(P) - r(P_1) - \cdots - r(P_k),
$$

$$(4.42)$$

$$
s(M) = r(P) - r(P_1) - \cdots - r(P_k).
$$

$$(4.43)$$

Proof: It is easily verified that both are orthogonal projectors.

$$
\frac{1}{k} \begin{bmatrix} P & P & \cdots & P \\ P & P & \cdots & P \\ \vdots & \vdots & \ddots & \vdots \\ P & P & \cdots & P \end{bmatrix}, \quad \begin{bmatrix} P_1 & 0 & \cdots & 0 \\ 0 & P_2 & \cdots & 0 \\ \vdots & \vdots & \ddots & \vdots \\ 0 & 0 & \cdots & P_k \end{bmatrix}
$$

Applying Theorem 3.2 to the difference of the two orthogonal projec-tors and simplifying, we obtain (4.40)–(4.43).

EXPANSION FORMULAS FOR INERTIAS OF ORTHOGONAL PROJECTORS ONTO RANGES OF BLOCK MATRICES

For the simplest row black matrix M = [A, B], the orthogonal projectors onto its range can be represented as

$$P_M = [A, B][A, B]^\dagger. \tag{5.1}$$

Hence, any formula for [A, B]$^+$ can be used to produce certain expression of P_M. Also, note that the column space of M is jointly spanned by the columns of A and B. Hence, the orthogonal projector P_M and the two orthogonal projectors P_A and P_B have some close links. One of the main concerns about (5.1) is to give its possible expansions or decompositions under various assumptions. Some previous work on this topic can be found, e.g., in [42]. In this section, we use the formulas for inertias of orthogonal projectors in the previous sections to derive a group of equalities for ranks/inertias of orthogonal projectors onto ranges of partitioned matrices.

Some simple results on the relations among the three orthogonal projectors are given below.

Lemma 5.1: Let $A \in \mathbb{C}^{m \times n}$, $B \in \mathbb{C}^{m \times k}$, and denote M = [A, B], $A_1 = E_B A$ and $B_1 = E_A B$. Then,

$$P_M P_A = P_A P_M = P_A, \qquad P_M P_B = P_B P_M = P_B, \tag{5.2}$$

$$\mathscr{R}(M) = \mathscr{R}(P_A + P_B), \tag{5.3}$$

$$r(P_M) = r(P_A) + r(P_{B_1}) = r(P_B) + r(P_{A_1}), \tag{5.4}$$

$$\mathscr{R}(M) = \mathscr{R}[A, B_1] = \mathscr{R}(P_A + P_{B_1}), \tag{5.5}$$

$$\mathscr{R}(M) = \mathscr{R}[A_1, B] = \mathscr{R}(P_{A_1} + P_B), \tag{5.6}$$

$$P_A P_{B_1} = P_{B_1} P_A = 0, \qquad P_{A_1} P_B = P_B P_{A_1} = 0. \tag{5.7}$$

Eq. (5.2) follows from the facts that $\mathfrak{R}(A) \subseteq \mathfrak{R}(M)$ and $\mathfrak{R}(B) \subseteq \mathfrak{R}(M)$; (5.3) follows from $\mathfrak{R}(P A + P B) \subseteq \mathfrak{R}(M)$ and $r(P_A + P_B) = r[P_{A'} P_B] = r(M)$; (5.4) follows from (1.2); (5.5) and (5.6) follow from and (5.7) follows from (1.1).

$$[A, B_1] = [A, B]\begin{bmatrix} I_n & -A^\dagger B \\ 0 & I_k \end{bmatrix}, \qquad [A_1, B] = [A, B]\begin{bmatrix} I_n & 0 \\ -B^\dagger A & I_k \end{bmatrix};$$

We first give some general results on the alternative expressions of P M in (5.1).

Theorem 5.2: Let A, B, M, A_1 and B1 be as given in Theorem $5._{1}$. Then, P M can be represented as

$$P_M = (AA^* + BB^*)(AA^* + BB^*)^\dagger, \tag{5.8}$$

$$P_M = (P_A + P_B)(P_A + P_B)^\dagger, \tag{5.9}$$

$$P_M = P_A + P_{B_1} = P_{A_1} + P_B. \tag{5.10}$$

Proof: Eq. (5.8) follows directly from the expansion of $M^\dagger = M* (MM*)^\dagger$. Eq. (5.9) follows from (5.3). Note that both $^{PA}+ P_{B1}$ and $P_{A1} + P_B$ are Hermitian. Also by (5.7),

$$(P_A + P_{B_1})^2 = P_A + P_{B_1}, \qquad (P_B + P_{A_1})^2 = P_B + P_{A_1}. \tag{5.11}$$

Thus, both $P_A + P_{B1}$ and $P_{A1} + P_B$ are orthogonal projectors. Combining this fact with (5.5) and (5.6) yields the unconditional decompositions in (5.10).

Applying Theorem 3.2 to the triple sides in (5.10) and simplifying, we obtain the following result. The details are omitted.

Theorem 5.3: Let A, B, M, A_1 and B_1 be as given in Theorem 5.1. Then

$$i_+(P_M - P_A) = r(P_M - P_A) = i_-(P_A - P_{B_1}) = i_-(P_{A_1} - P_{B_1}) = r(M) - r(A), \tag{5.12}$$

$$i_+(P_M - P_{B_1}) = r(P_M - P_{B_1}) = i_+(P_A - P_{B_1}) = r(A), \tag{5.13}$$

$$i_+(P_M - P_B) = r(P_M - P_B) = i_-(P_B - P_{A_1}) = i_-(P_{A_1} - P_{B_1}) = r(M) - r(B), \tag{5.14}$$

$$i_+(P_M - P_{A_1}) = r(P_M - P_{A_1}) = i_+(P_B - P_{A_1}) = r(B), \tag{5.15}$$

$$i_+(P_M - P_{A_1} - P_{B_1}) = i_+(P_A - P_{A_1}) = i_+(P_B - P_{B_1}) = r(A^*B), \tag{5.16}$$

$$i_-(P_M - P_{A_1} - P_{B_1}) = i_-(P_A - P_{A_1}) = i_-(P_B - P_{B_1}) = r(M) + r(A^*B) - r(A) - r(B). \tag{5.17}$$

Hence,

a) $P_M \le P_A \Leftrightarrow P_M = P_A \Leftrightarrow P_A \ge P_{B_1} \Leftrightarrow P_{A_1} \ge P_{B_1} \Leftrightarrow \Re(B) \subseteq \Re(A)$.

b) $P_M \le P_B \Leftrightarrow P_M = P_B \Leftrightarrow P_B \ge P_{A_1} \Leftrightarrow P_{B_1} \ge P_{A_1} \Leftrightarrow \Re(A) \subseteq \Re(B)$.

c) $P_M \le P_{A_1} + P_{B_1} \Leftrightarrow P_A \le P_{A_1} \Leftrightarrow P_B \le P_{B_1} \Leftrightarrow A*B = 0$.

d) $P_M \ge P A1 + P B1 \Leftrightarrow P A \ge P A1 \Leftrightarrow P B \ge P B1 \Leftrightarrow r(M) = r(A) + r(B) - r(A*B)$.

Since the Moore–Penrose inverse of M in (5.1) can be written in different forms under various assumptions. Correspondingly, the P M in (5.1) can be represented in some particular forms. For instance, it is easily derived from (1.1) that $A*B = 0 \Leftrightarrow B* A = 0 \Leftrightarrow A^\dagger B = 0 \Leftrightarrow B^\dagger A = 0$. A well-known result associated with this equivalence is

$$[A, B]^\dagger = \begin{bmatrix} A^\dagger \\ B^\dagger \end{bmatrix} \quad \Leftrightarrow \quad A^*B = 0. \tag{5.18}$$

Moreover, the present author showed in [48, 49] that

$$r\left([A, B]^\dagger - \begin{bmatrix} A^\dagger \\ B^\dagger \end{bmatrix}\right) = r[BB^*A, AA^*B]. \tag{5.19}$$

Hence, the equivalence in (5.18) is a direct consequence of (5.19). Under (5.18), the orthogonal projector in (5.1) can be rewritten as the sum of two orthogonal projectors

$$P_M = [A, B][A, B]^\dagger = AA^\dagger + BB^\dagger = P_A + P_B.$$
(5.20)

Eqs. (5.19) and (5.20) prompt us to obtain the following result on the difference of both sides of (5.20).

Theorem 5.4: Let $A \in \mathbb{C}^{m \times n}$ $B \in \mathbb{C}^{m \times k}$, and denote M = [A, B]. Then,

$$i_+(P_M - P_A - P_B) = r(M) + r(A^*B) - r(A) - r(B),$$
(5.21)

$$i_-(P_M - P_A - P_B) = r(A^*B),$$
(5.22)

$$r(P_M - P_A - P_B) = r(M) + 2r(A^*B) - r(A) - r(B),$$
(5.23)

$$s(P_M - P_A - P_B) = r(M) - r(A) - r(B).$$
(5.24)

Hence,

a) $P_M \geq P_A + P B \Leftrightarrow P_M = P_A + P_B \Leftrightarrow A*B = 0.$
b) $P_M \leq P_A + P_B$ if and only if r(M) = r(A) + r(B) − r(A*B).
c) The signature of $P_M - P_A - P_B$ is zero if and only if r (M) = r (A) + r(B), i.e., $\mathfrak{R}$ (A) ∩ $\mathfrak{R}$ (B) = {0}.

Proof: It is obvious that $\mathfrak{R}$ (P$_A$) ⊆ $\mathfrak{R}$ (P$_M$), R(P B) ⊆ $\mathfrak{R}$ (P$_M$), and

$$r(P_M) = r(M), \qquad r(P_A) = r(A), \qquad r(P_B) = r(B), \qquad r(P_A P_B) = r(A^*B).$$
(5.25)

Then, applying (4.11), (4.12), (4.15) and (5.25) to $P_M - P_A - P_B$ yields (5.20)–(5.24). Results (a)–(c) follow from (5.20)–(5.24) and Lemma 1.1.

Other formulas for the ranks/inertias of the orthogonal projectors onto A, B and [A, B] are given below.

Theorem 5.5: Let A, B, and M is as given in Theorem 5.1. Also, assume that C is a matrix such that $\mathfrak{R}$ (C) = $\mathfrak{R}$ (A) ∩ $\mathfrak{R}$ (B). Then,

$$i_+(2P_M - P_A - P_B) = r(2P_M - P_A - P_B) = 2r(M) - r(A) - r(B),$$
(5.26)

$$i_+(P_M - P_C) = r(P_M - P_C) = 2r(M) - r(A) - r(B). \tag{5.27}$$

Hence,

$$2P_M \leqslant P_A + P_B \quad \Leftrightarrow \quad 2P_M = P_A + P_B \quad \Leftrightarrow \quad P_M \leqslant P_C \quad \Leftrightarrow \quad P_M = P_C \quad \Leftrightarrow \quad \mathscr{R}(A) = \mathscr{R}(B).$$
$$\tag{5.28}$$

Proof: It follows from Theorems 3.2 and 4.1(b).

The following results can easily be derived from Theorem 3.2, and the details are also omitted.

Theorem 5.6: Let $A \in \mathbb{C}^{m \times n}$, $C \in \mathbb{C}^{l \times n}$, and denote

$$M = \begin{bmatrix} A \\ C \end{bmatrix}, \qquad N = \begin{bmatrix} A & 0 \\ 0 & C \end{bmatrix}.$$

Then,

$$i_-(P_M - P_N) = r(P_M - P_N) = r(A) + r(C) - r(M). \tag{5.29}$$

Hence,

$$P_M \geqslant P_N \quad \Leftrightarrow \quad P_M = P_N \quad \Leftrightarrow \quad r(M) = r(A) + r(C). \tag{5.30}$$

Theorem 5.7: Let $A \in \mathbb{C}^{m \times n}$, $B \in \mathbb{C}^{m \times k}$, and $D \in \mathbb{C}^{l \times n}$, and denote

$$M = \begin{bmatrix} A & B \\ 0 & D \end{bmatrix}, \qquad N = \begin{bmatrix} A & 0 \\ 0 & D \end{bmatrix}.$$

Then,

$$i_+(P_M - P_N) = r[A, B] - r(A), \tag{5.31}$$

$$i_-(P_M - P_N) = r[A, B] + r(D) - r(M), \tag{5.32}$$

$$r(P_M - P_N) = 2r[A, B] + r(D) - r(M) - r(A),$$

$$(5.33)$$

$$s(P_M - P_N) = r(M) - r(A) - r(D).$$

$$(5.34)$$

Hence,

a) $P_M - P_N \geq 0$ if and only if r (M) = r [A, B] + r (D).

b) $P_M - P_N \leq 0$ if and only if $\mathcal{R}$ (B) $\subseteq$ $\mathcal{R}$ (A).

c) $P_M = P_N \Leftrightarrow \mathcal{R}$ (B) $\subseteq$ $\mathcal{R}$ (A) and r(M) = r(A) + r(D).

d) The signature of $P_M - P_N$ is zero if and only if r (M) = r (A) + r(D).

Theorem 5.8: Let A $\in \mathbb{C}^{m \times k}$, B $\in \mathbb{C}^{m \times k}$, C $\in \mathbb{C}^{l \times n}$ and D $\in \mathbb{C}^{l \times k}$, and denote

$$M = \begin{bmatrix} A & B \\ C & D \end{bmatrix}, \qquad N = \begin{bmatrix} A & 0 \\ 0 & D \end{bmatrix}.$$

Then,

$$i_+(P_M - P_N) = r[A, B] + r[C, D] - r(A) - r(D),$$

$$(5.35)$$

$$i_-(P_M - P_N) = r[A, B] + r[C, D] - r(M),$$

$$(5.36)$$

$$r(P_M - P_N) = 2r[A, B] + 2r[C, D] - r(M) - r(A) - r(D),$$

$$(5.37)$$

$$s(P_M - P_N) = r(M) - r(A) - r(D).$$

$$(5.38)$$

Hence,

a) $P_M - P_N \geq 0$ if and only if r (M) = r [A, B] + r[C, D].

b) $P_M - P_N \leq 0 \Leftrightarrow \mathcal{R}$ (B) $\subseteq$ $\mathcal{R}$ (A) and $\mathcal{R}$ (C) $\subseteq$ $\mathcal{R}$ (D).

c) $P_M = P_N$ if and only if r (M) = r (A) + r (D), $\mathcal{R}$ (B) $\subseteq$ $\mathcal{R}$ (A) and $\mathcal{R}$ (C) $\subseteq$ R(D).

d) The signature of $P_M - P_N$ is zero if and only if r (M) = r (A) + r (D).

e) If M $\geq$ 0, then $P_M \leq P_N$.

More expressions consisting of the orthogonal projectors onto the block matrix M = [A, B] and its sub matrices can be formulated, such as,

$$P_M - 2^{-1}P_A P_B - 2^{-1}P_B P_A, \qquad P_M - P_A - P_B + 2^{-1}P_A P_B + 2^{-1}P_B P_A,$$
$$P_M - 2P_A(P_A + P_B)^\dagger P_B, \qquad P_M - P_A - P_B + 2P_A(P_A + P_B)^\dagger P_B.$$

It is also of interest to derive possible closed-form formulas for the ranks/inertias of these matrix expressions.

EXPANSION FORMULAS FOR INERTIAS OF HERMITIAN UNITARY MATRICES AND THEIR OPERATIONS

It is easy to verify that if $A \in \mathbb{C}_{HU}^m$, then its transformations $P = (I_m \pm A)/2$ satisfy $P^2 = P = P*$, namely, the two matrices P are orthogonal projectors. In view of this fact, the formulas for inertias of orthogonal projectors in the previous sections can be extended to Hermitian unitary matrices through the transformations. For instance, if both A and B are Hermitian unitary matrices of the same size, then their transformations $P = (I_m \pm A)/2$ and $Q = (I_m \pm B)/2$ are orthogonal projectors, and the matrix pencil $\lambda I_m + \lambda_1 A + \lambda_2 B$ can be represented as

$$\lambda I_m + \lambda_1 A + \lambda_2 B = (\lambda - \lambda_1 - \lambda_2)I_m + 2\lambda_1 P_1 + 2\lambda_2 Q_1, \qquad (6.1)$$

$$\lambda I_m + \lambda_1 A + \lambda_2 B = (\lambda + \lambda_1 + \lambda_2)I_m - 2\lambda_1 P_2 - 2\lambda_2 Q_2, \qquad (6.2)$$

Where

$$P_1 = (I_m + A)/2, \qquad Q_1 = (I_m + B)/2, \qquad P_2 = (I_m - A)/2, \qquad Q_2 = (I_m - B)/2$$

Are four orthogonal projectors. In particular, we have

$$A + B = 2I_m - 2P_1 - 2Q_1 = -2I_m + 2P_2 + 2Q_2. \qquad (6.3)$$

Applying Corollary 4.2 to (6.3) yields the following result.

Theorem 6.1: Let $A, B \in \mathbb{C}_{HU}^m$. Then

$$i_+(A+B) = r\big[(I_m + A)(I_m + B)\big] = 2^{-1}\operatorname{tr}(A+B) + r\big[(I_m - A)(I_m - B)\big], \tag{6.4}$$

$$i_-(A+B) = -2^{-1}\operatorname{tr}(A+B) + r\big[(I_m + A)(I_m + B)\big] = r\big[(I_m - A)(I_m - B)\big], \tag{6.5}$$

$$r(A+B) = r\big[(I_m + A)(I_m + B)\big] + r\big[(I_m - A)(I_m - B)\big], \tag{6.6}$$

$$s(A+B) = 2^{-1}\operatorname{tr}(A+B). \tag{6.7}$$

Hence,

a) $A + B > 0$ if and only if $A = B = I_m$.
b) $A + B < 0$ if only if $A = B = -I_m$.
c) $A + B \geq 0$ if and only if $(I_m - A)(I_m - B) = 0$.
d) $A + B \leq 0$ if and only if $(I_m + A)(I_m + B) = 0$.

Proof: Applying (4.20) to (6.1) and simplifying, we obtain

$$
\begin{aligned}
i_\pm(\lambda_1 A + \lambda_2 B) &= i_\pm\big[-(\lambda_1 + \lambda_2)I_m + 2\lambda_1 P_1 + 2\lambda_2 Q_1\big] \\
&= i_\pm\big[-(\lambda_1 + \lambda_2)I_m\big] - i_\mp(2\lambda_1 P_1) - i_\mp(2\lambda_2 Q_1) \\
&\quad + i_\mp \begin{bmatrix} [(2\lambda_1)^{-1} - (\lambda_1 + \lambda_2)^{-1}]P_1 & -(\lambda_1 + \lambda_2)^{-1}P_1 Q_1 \\ -(\lambda_1 + \lambda_2)^{-1}Q_1 P_1 & [(2\lambda_2)^{-1} - (\lambda_1 + \lambda_2)^{-1}]Q_1 \end{bmatrix} \\
&= i_\mp\big[(\lambda_1 + \lambda_2)I_m\big] - i_\mp\big[\lambda_1(I_m + A)\big] - i_\mp\big[\lambda_2(I_m + B)\big] \\
&\quad + i_\mp \begin{bmatrix} [(2\lambda_1)^{-1} - (\lambda_1 + \lambda_2)^{-1}](I_m + A) & -(\lambda_1 + \lambda_2)^{-1}(I_m + A)(I_m + B) \\ -(\lambda_1 + \lambda_2)^{-1}(I_m + B)(I_m + A) & [(2\lambda_2)^{-1} - (\lambda_1 + \lambda_2)^{-1}](I_m + B) \end{bmatrix}.
\end{aligned}
\tag{6.8}
$$

Applying (4.20) to (6.2) and simplifying, we obtain

$$
\begin{aligned}
i_\pm(\lambda_1 A + \lambda_2 B) &= i_\pm\big[(\lambda_1 + \lambda_2)I_m\big] - i_\pm\big[\lambda_1(I_m - A)\big] - i_\pm\big[\lambda_2(I_m - B)\big] \\
&\quad + i_\mp \begin{bmatrix} [(\lambda_1 + \lambda_2)^{-1} - (2\lambda_1)^{-1}](I_m - A) & (\lambda_1 + \lambda_2)^{-1}(I_m - A)(I_m - B) \\ (\lambda_1 + \lambda_2)^{-1}(I_m - B)(I_m - A) & [(\lambda_1 + \lambda_2)^{-1} - (2\lambda_2)^{-1}](I_m - B) \end{bmatrix}.
\end{aligned}
\tag{6.9}
$$

Setting $\lambda_1 = \lambda_2 = 1$ in (6.8) and (6.9) leads to

$$i_{\pm}(A + B) = i_{\mp}(I_m) - i_{\mp}(I_m + A) - i_{\mp}(I_m + B) + r\left[(I_m + A)(I_m + B)\right],$$

(6.10)

$$i_{\pm}(A + B) = i_{\pm}(I_m) - i_{\pm}(I_m - A) - i_{\pm}(I_m - B) + r\left[(I_m - A)(I_m - B)\right].$$

(6.11)

Also, note that

$i_+(I_m) = m, \qquad i_-(I_m) = 0,$

$i_+(I_m + A) = 2^{-1}\,\mathrm{tr}(I_m + A), \qquad i_-(I_m + A) = 0, \qquad i_+(I_m - A) = 2^{-1}\,\mathrm{tr}(I_m - A), \qquad i_-(I_m - A) = 0,$

$i_+(I_m + B) = 2^{-1}\,\mathrm{tr}(I_m + B), \qquad i_-(I_m + B) = 0, \qquad i_+(I_m - B) = 2^{-1}\,\mathrm{tr}(I_m - B), \qquad i_-(I_m - B) = 0.$

Therefore, (6.10) and (6.11) reduce to (6.4) and (6.5). Eqs. (6.6) and (6.7) follow from (6.4) and (6.5). Results (a)–(d) follow (6.4) and (6.5) and Lemma 1.1.

More formulas for the partial inertias of two Hermitian unitary matrices and their operations can be derived. For instance, applying Theorem 2.1 to the sum of two Hermitian unitary matrices $A, B \in \mathbb{C}^{m \times m}$ leads to

$$i_{\pm}\left[(A + B) - (A + B)^2\right] = i_{\pm}(A + B) + i_{\pm}(I_m - A - B) - i_{\pm}(I_m),$$

(6.12)

$$i_{\pm}\left[I_m - (A + B)^2\right] = i_{\pm}(I_m + A + B) + i_{\pm}(I_m - A - B) - i_{\pm}(I_m),$$

$$i_{\pm}(AB + BA) = i_{\mp}\left[I_m - (A/\sqrt{2} + B/\sqrt{2})^2\right]$$

(6.13)

$$= i_{\mp}(\sqrt{2}I_m + A + B) + i_{\mp}(\sqrt{2}I_m - A - B) - i_{\mp}(I_m).$$

(6.14)

Note that $I_m \pm (A + B)$ and $\sqrt{2}I_m \pm (A + B)$ can be rewritten as

$$I_m + A + B = 3I_m - 2P_1 - 2Q_1 = -I_m + 2P_2 + 2Q_2,$$

(6.15)

$$I_m - A - B = I_m + 2P_1 + 2Q_1 = -3I_m - 2P_2 - 2Q_2,$$

(6.16)

$$\sqrt{2}I_m + A + B = (\sqrt{2} + 2)I_m - 2P_1 - 2Q_1 = (\sqrt{2} - 2)I_m + 2P_2 + 2Q_2,$$

(6.17)

$$\sqrt{2}I_m - A - B = (\sqrt{2} - 2)I_m + 2P_1 + 2Q_1 = (\sqrt{2} + 2)I_m - 2P_2 - 2Q_2, \qquad (6.18)$$

Where $P_1 = (I_m + A)/2$, $Q_1 = (I_m + B)/2$, $P_2 = (I_m - A)/2$ and $Q_2 = (I_m - B)/2$ are orthogonal projectors. In these cases, applying Corollary 4.2 to (6.15)–(6.18) may yield some expansion formulas for the ranks/inertias of $I_m \pm (A + B)$ and $\sqrt{2}I_m \pm (A + B)$. Substituting these formulas into (6.12)–(6.14) may also yield some expansion formulas for the ranks/inertias of $(A + B) - (A + B)^2$, $I_m - (A + B)^2$ and $A B + B A$.

CONCLUDING REMARKS

In this paper, we constructed some congruence transformations for block Hermitian matrices consisting of Hermitian matrices, orthogonal projectors and their operations. From these Hermitian congruence transformations and the well-known Sylvester's law of inertia, we obtained a variety of explicit expansion formulas for ranks/inertias of Hermitian matrix polynomials, orthogonal projectors, Hermitian unitary matrices and their operations. Using these formulas, we further characterized many equalities and inequalities for Hermitian matrix polynomials and orthogonal projectors in the Löwner partial ordering. The algebraic methods adopted in the manipulations are quite elementary, and the results obtained seem quite simple and interesting. Therefore, the investigation in this paper would bring us deeper understanding to properties of Hermitian matrix polynomials, orthogonal projectors, Hermitian unitary matrices and their operations.

In addition to the simple matrix expressions considered in the previous sections, various general expressions consisting of orthogonal projectors may occur in matrix theory and applications. These expressions can, in general, be written as

$$p(P_1, P_2, \ldots, P_k), \qquad (7.1)$$

Where P_1, P_2,..., P_k are a group of orthogonal projectors of appropriate sizes. If the expression is Hermitian, it would also be of interest to establish some $*$-congruent block transformation equalities associated with (7.1), and then to establish expansion formulas for the inertia/rank of the matrix expression. In particular, motivated by (3.12), (3.13), (4.3) and (4.4), a challenging task is to establish expansion formulas for the ranks/inertias of some general Hermitian matrix pencils, such as, $\lambda_1 P_1 + \lambda_2 P_2 + \lambda_3 P_3$, where $P_1, P_2, P_3 \in \mathbb{C}_{OP}^m$, and $\lambda_1, \lambda_2, \lambda_3$ are nonzero real numbers, as well as, $\lambda P + \lambda_1 P_1 + \cdots + \lambda_k P_k$, where $P, P_i \in \mathbb{C}_{OP}^m$ satisfy $\mathfrak{R}(P_i) \subseteq \mathfrak{R}(P)$, $i = 1, \ldots, k$, and $\lambda, \lambda_1, \ldots, \lambda_k$ are nonzero real numbers.

Also, we point out that the congruence transformations in the previous sections for Hermitian matrices and orthogonal projectors also hold in general frames, such as, Hermitian operators and orthogonal projectors in a Hilbert space. In this event, the inertias of self-adjoint operators can accordingly be defined; see, e.g., [18, 32]. In addition, orthogonal projectors can also be defined over rings with involution through the products of elements with their Moore–Penrose inverses; see [41]. In this case, it would be of interest to consider extensions of the work in this paper to orthogonal projectors over rings with involution

As is known to all, the rank and inertia of a matrix are two basic concepts in elementary linear algebra. Any results on ranks/inertias of matrices, in particular, various closed-form formulas for ranks/inertias of matrices, are easy to understand within the scope of common knowledge in linear algebra. The conventional tools for handling ranks/inertias of matrices symbolically, as demonstrated in deriving the formulas in the previous sections, are nothing but the usual elementary operations and congruence transformations for matrices. In the past two decades, the present author has been devoting on this topic, and has proved a huge amount of results on ranks/inertias of matrices and their applications by the elementary methods mentioned above. It is expected that more and more results on ranks/inertias of matrices can be discovered, which, I believe, will become a part of core contents in linear algebra.

ACKNOWLEDGMENTS

The author thanks Professors Ying Li, Zhongpeng Yang and Fuzhen Zhang for their helpful comments on this paper. The author is grateful to the referee for constructive comments and suggestions on this paper.

REFERENCES

1. T.W. Anderson, G.P.H. Styan, Cochran's theorem, rank additivity and tripotent matrices, in: G. Kallianpur, P.R. Krishnaiah, J.K. Ghosh (Eds.), Statistics and Probability: Essays in Honor of C.R. Rao, North-Holland, 1982, pp. 1–23.

2. Y.-H. Au-Yeung, On the semi-definiteness of the real pencil of two Hermitian matrices, Linear Algebra Appl. 10 (1975) 71–76.

3. J.K. Baksalary, O.M. Baksalary, Idempotency of linear combinations of two idempotent matrices, Linear Algebra Appl. 321 (2000) 3–7.

4. J.K. Baksalary, O.M. Baksalary, When is a linear combination of two idempotent matrices the group involutory matrix?, Linear Multilinear Algebra 54 (2006) 429–435.

5. J.K. Baksalary, O.M. Baksalary, H. Özdemir, A note on linear combinations of commuting tripotent matrices, Linear Algebra Appl. 388 (2004) 45–51.

6. J.K. Baksalary, O.M. Baksalary, T. Szulc, A property of orthogonal projectors, Linear Algebra Appl. 354 (2002) 35–39.

7. O.M. Baksalary, Idempotency of linear combinations of three idempotent matrices, two of which are disjoint, Linear Algebra Appl. 388 (2004) 67–78.

8. O.M. Baksalary, J. Benitez, Idempotency of linear combinations of three idempotent matrices, two of which are commuting, Linear Algebra Appl. 424 ´ (2007) 320–337.

9. W. Barrett, H.T. Hall, R. Loewy, The inverse inertia problem for graphs: cut vertices, trees, and a counterexample, Linear Algebra Appl. 431 (2009) 1147–1191.

10. J. Benítez, V. Rakocevi ˇ c, Applications of CS decomposition in linear combinations of two orthogonal projectors, Appl. Math. Comput. 203 (2008) 761– ´ 769.

11. J. Benítez, V. Rakocevi ˇ c, On the spectrum of linear combinations of two projections in ´ C∗-algebras, Linear Multilinear Algebra 58 (2010) 673–679.

12. A. Ben-Israel, T.N.E. Greville, Generalized Inverses: Theory and Applications, second ed., Springer, New York, 2003.

13. D.S. Bernstein, Matrix Mathematics: Theory, Facts and Formulas, second ed., Princeton University Press, Princeton, 2009.

14. J. Bérubé, R.E. Hartwig, G.P.H. Styan, On canonical correlations and the degrees of non-orthogonality in the three-way layout, in: K. Matusita, M.L. Puri, T. Hayakawa (Eds.), Statistical Sciences and Data Analysis: Proceedings of the Third Pacific Area Statistical Conference, Makuhari (Chiba, Tokyo), Japan, December 11–13, 1991, VSP International Science Publishers, Utrecht, The Netherlands, 1993, pp. 247–252.

15. P. Binding, The inertia of a Hermitian pencil, Linear Algebra Appl. 63 (1984) 179–191.

16. A. Björck, Numerical Methods for Least Squares Problems, SIAM, Philadelphia, PA, 1996.

17. P.E. Bjørstad, J. Mandel, On the spectra of sums of orthogonal projections with applications to parallel computing, BIT 31 (1991) 76–88.

18. B.E. Cain, An inertia theory for operators on a Hilbert space, J. Math. Anal. Appl. 41 (1973) 97–114.

19. C.R. Crawford, Y.S. Moon, Finding a positive definite linear combination of two Hermitian matrices, Linear Algebra Appl. 51 (1983) 37–48.

20. D. Hershkowitz, H. Schneider, On the inertia of intervals of matrices, SIAM J. Matrix Anal. Appl. 11 (1990) 565–574.

21. A. Galántai, Projectors and Projection Methods, Kluwer Academic Publishers, 2004.

22. J. Groß, On the product of orthogonal projectors, Linear Algebra Appl. 289 (1999) 141–150.

23. J. Groß, G. Trenkler, Problem 96.4.3: orthogonal projectors, Econometric Theory 12 (1996) 744; solved by S. Puntanen, G.P.H. Styan, Econometric Theory 13 (1997) 764–765.

24. G.H. Golub, C.F. Van Loan, Matrix Computations, third ed., Johns Hopkins Studies in the Mathematical Sciences, Johns Hopkins University Press, Baltimore, MD, 1996.

25. E.V. Haynsworth, Determination of the inertia of a partitioned Hermitian matrix, Linear Algebra Appl. 1 (1968) 73–81.

26. E.V. Haynsworth, A.M. Ostrowski, On the inertia of some classes of partitioned matrices, Linear Algebra Appl. 1 (1968) 299–316.

27. [27] T.M. Hoang, T. Thierauf, The complexity of the inertia, Lecture Notes in Comput. Sci. 2556 (2002) 206–217.

28. T.M. Hoang, T. Thierauf, The complexity of the inertia and some closure properties of GapL, in: Proceedings of the Twentieth Annual IEEE Conference on Computational Complexity, 2005, pp. 28–37.

29. L. Hogben, Handbook of Linear Algebra, Chapman & Hall/CRC, 2007.

30. R.A. Horn, C.R. Johnson, Matrix Analysis, Cambridge University Press, Cambridge, 1985.

31. Kh.D. Ikramov, Matrix pencils: theory, applications, and numerical methods, J. Math. Sci. 64 (1993) 783–853.

32. C.R. Johnson, M. Lundquist, Operator matrices with chordal inverse patterns, Oper. Theory Adv. Appl. 59 (1992) 234–251.

33. H.Th. Jongen, T. Möbert, J. Rückmann, K. Tammer, On inertia and Schur complement in optimization, Linear Algebra Appl. 95 (1987) 97–109.

34. N.Ya. Krupnik, A.S. Markus, I.A. Fel'dman, Norm of a linear combination of projectors in Hilbert space, Funct. Anal. Appl. 23 (1989) 327–329.

35. K. Löwner, Über monotone matrix funktionen, Math. Z. 38 (1934) 177–216.

36. G. Marsaglia, G.P.H. Styan, Equalities and inequalities for ranks of matrices, Linear Multilinear Algebra 2 (1974) 269–292.

37. A.W. Marshall, I. Olkin, Inequalities: Theory of Majorization and Its Applications, Academic Press, Orlando, FL, 1979.

38. L. Mirsky, An Introduction to Linear Algebra, second corrected reprint edition, Dover, New York, 1990.

39. Y. Nakamura, Any Hermitian matrix is a linear combination of four projections, Linear Algebra Appl. 61 (1984) 133–139.

40. H. Özdemir, A.Y. Özban, On idempotency of linear combinations of idempotent matrices, Appl. Math. Comput. 159 (2004) 439–448.

41. P. Patricio, C.M. Araújo, Moore–Penrose invertibility in involutory rings: the case aa† = bb† , Linear Multilinear Algebra 58 (2010) 445–452.

42. C.R. Rao, H. Yanai, General definition and decomposition of projectors and some applications to statistical problems, J. Statist. Plann. Inference 3 (1979) 1–17.

43. I.M. Spitkovsky, Once more on algebras generated by two projections, Linear Algebra Appl. 208/209 (1994) 377–395.

44. I.M. Spitkovsky, On polynomials in two projections, Electron. J. Linear Algebra 15 (2006) 154–158.

45. J.J. Sylvester, A demonstration of the theorem that every homogeneous quadratic polynomial is reducible by real orthogonal substitutions to the form of a sum of positive and negative squares, Lond. Edinb. Dublin Philos. Magazine J. Sci., Fourth Ser. 4 (1852) 138–142; reprinted in: H.F. Baker (Ed.), The Collected Mathematical Papers of James Joseph Sylvester, I, Cambridge University Press, 1904, pp. 378–381.

46. R.C. Thompson, Pencils of complex and real symmetric and skew matrices, Linear Algebra Appl. 147 (1991) 323–371.

47. R.C. Thompson, Root spreads for polynomials and Hermitian matrix pencils, Linear Algebra Appl. 220 (1995) 419–433.

48. Y. Tian, Rank equalities for block matrices and their Moore–Penrose inverses, Houston J. Math. 30 (2004) 483–510.

49. Y. Tian, Special forms of generalized inverses of row block matrices, Electron. J. Linear Algebra 13 (2005) 249–261.

50. Y. Tian, Equalities and inequalities for inertias of Hermitian matrices with applications, Linear Algebra Appl. 433 (2010) 263–296.

51. Y. Tian, Rank and inertia of submatrices of the Moore–Penrose inverse of a Hermitian matrix, Electron. J. Linear Algebra 20 (2010) 226–240.

52. Y. Tian, Equalities for orthogonal projectors and their operations, Cent. Eur. J. Math. 8 (2010) 855–870.

53. Y. Tian, Two universal similarity factorization equalities for commutative involutory and idempotent matrices and their applications, Linear Multilinear Algebra, in press.

54. Y. Tian, Completing block Hermitian matrices with maximal and minimal ranks and inertias, Electron. J. Linear Algebra, in press.

55. Y. Tian, G.P.H. Styan, Rank equalities for idempotent and involutory matrices, Linear Algebra Appl. 335 (2001) 101–117.

56. Y. Tian, G.P.H. Styan, How to establish universal block-matrix factorizations, Electron. J. Linear Algebra 8 (2001) 115–127.

57. Y. Tian, G.P.H. Styan, Rank equalities for idempotent matrices with applications, J. Comput. Appl. Math. 191 (2006) 77–97.

58. N.-K. Tsing, F. Uhlig, Inertia, numerical range, and zeros of quadratic forms for matrix pencils, SIAM J. Matrix Anal. Appl. 12 (1991) 146–159.

59. F. Uhlig, Computing the inertias in symmetric matrix pencils, Linear Algebra Appl. 201 (1994) 199–209.

60. H. Väliaho, Determining the inertia of a matrix pencil as a function of the parameter, Linear Algebra Appl. 106 (1988) 245–258.

61. N. Vasilevsky, I.M. Spitkovsky, On the algebra generated by two projections, Dokl. Akad. Nauk Ukrain. SSR Ser. A 8 (1981) 10–13.

62. S. Zlobec, An explicit form of the Moore–Penrose inverse of an arbitrary complex matrix, SIAM Rev. 12 (1970) 132–134.

CITATION

Yongge Tian, Expansion formulas for the inertias of Hermitian matrix polynomials and matrix pencils of orthogonal projectors, Journal of Mathematical Analysis and Applications, Volume 376, Issue 1, 1 April 2011, Pages 162-186, ISSN 0022-247X, http://dx.doi.org/10.1016/j.jmaa.2010.09.038.

Equivalent Conditions for Noncentral Generalized Laplacianness and Independence of Matrix Quadratic Forms

Jianhua Hu[1]
School of Statistics and Management,
Shanghai University of Finance and
Economics, 777 Guoding Rd., Shanghai
200433, PR China

8

ABSTRACT

Let Y be an n × p multivariate normal random matrix with general covariance Σ_Y and W be a symmetric matrix. In the present article, the property that a matrix quadratic form Y'WY is distributed as a difference of two independent (no central) Wish art random matrices is called the (no central) generalized Laplacianness (GL). Then a set of algebraic results are obtained which will give the necessary and sufficient conditions for the (no central) GL of a matrix quadratic form. Further, two extensions of Cochran's theorem concerning the (no central) GL and independence of a family of matrix quadratic forms are developed.

INTRODUCTION

In the research of the distribution of quadratic forms, the problem that a quadratic form is distributed as a difference of two independent chi-squire random variables and its generalization have been investigated by many scholars. Usually, the equivalent algebraic conditions are expected to characterize the property that a quadratic form is distributed as a difference of two independent chi-squire random variables.

Luther [9] established the equivalence between the distribution of a quadratic form as the unique difference of two stochastically independent chi-square distributions and the tripotency of its underlying matrix. Later, Baldessari [2] developed necessary and sufficient conditions under which a quadratic form, in normal random variables, is distributed as a given linear combination of independent chi-square random variables, generalizing Luther's result. Tan [22] extended Baldessari's results to a quadratic form in possibly singular normal random variables. Khatri [8] further extended Baldessari's result to the singular covariance matrix, to quadratic forms and to quadratic expressions. Moreover, Tan [21] extended the problem from the unilabiate case to the multivariate case and obtained some extensions of Cochran's theorem concerning differences of independent noncentral Wishart random matrices, where the covariance of normal random matrix Y is the structure of Kronecker product $A \otimes \Sigma$. Wong and Wang [24] extended Tan's results to the case of a general covariance matrix, meaning that the collection of all np elements in Y has an arbitrary np × np covariance matrix. Masaro and Wong [11] derived a set of necessary and sufficient conditions for Laplace–Wishart distribution associated with matrix quadratic forms when Y follows a multivariate normal distribution with zero mean and Laplace–Wishart distribution has a diagonal covariance. Brief summaries of the related development are available in Anderson and Styan [1] and Hu [6].

Other scholars who also worked on chi-square difference and their generalizations, in distribution function approach, include Pearson et al. [15], Gurland [5], Shah [20], Robinson [19], Press [16] and Provost [17]. The similar research also appeared for gamma difference, see Mathai [13]. Cochran's theorem was proposed in Cochran [3]. A summary of the extensions of Cochran's theorem concerning chi-squareness or Wishartness and independence is given in Hu [6, 7], and recently Masaro and Wong [12].

This article will extend Tan's results to the general covariance ΣY of Y as did in Wong and Wang [24]. The new results obtained in this article, based on Masaro and Wong's work [11] greatly improve Wong and Wang's works as well as extend Masaro and Wong's work.

In this article, Y denotes an n × p multivariate normal random matrix with general covariance ΣY and W denotes a symmetric matrix. The property that a matrix quadratic form Y'WY is distributed as a difference of two independent (noncentral) Wishart random matrices is called the (noncentral) generalized Laplacianness (GL). The terminology is quoted from Mathai [13]. The organization of this article is as follows.

In Section 2, some necessary preliminaries are summarized. Conditions for the GL of a matrix quadratic form are established in Section 3 and a general extension of Cochran's theorem concerning the GL and independence of a family of matrix quadratic forms is developed in Section 4. The parallel results to, respectively, the noncentral GL of a matrix quadratic form and an extension of Cochran's theorem concerning the noncentral GL and independence of a family of matrix quadratic forms are established in Sections 5 and 6. The concluding remarks is briefly stated in Section 7. The related lemmas are presented in Appendix.

PRELIMINARIES

In this paper, $\mathcal{M}_{n \times p}$ denotes the set of n × p matrices over the real set $\mathfrak{R}$. The trace inner product $\langle , \rangle$ equipped in $\mathcal{M}_{n \times p}$ is defined as $\langle A, B \rangle = \mathrm{tr}(AB')$ for all A, B $\in \mathcal{M}_{n \times p}$, where B' is the transpose of B. $\| \ \|$ denotes the trace norm in $\mathcal{M}_{n \times p}$ and $|A|$ denotes the determinant of A. $\mathcal{S}_p$ denotes the set of symmetric matrices of order p over the real set $\mathfrak{R}$. $\mathcal{N}_p$ denotes the set of nonnegative definite matrices of order p over the real set $\mathfrak{R}$. I_m denotes the identity matrix of order m.

We use e_{ij} to denote the matrix whose ij^{th} entry is 1 and all other entries 0 and E_{ij} the symmetric matrix of order p whose ij^{th} entry and ji^{th} entry both are 1 and all other entries 0. Write $\mathcal{B}_p = \{E_{ij}: 1 \leq i \leq j \leq p\}$, called the basic base of the set $\mathcal{S}_p$.

For a nonnegative definite matrix Σ of order p, there exists an orthogonal matrix H such that H'ΣH = diag $[\Sigma_1, \Sigma_2,..., \Sigma_p]$ with $\Sigma_i \geq 0$. Write $\mathcal{H}_p$

$= \{HE_{ij}H' : 1 \le i \le j \le p, E_{ij} \in \mathcal{B}_p\}$, called the similar base, associated with Σ, of the set $\mathcal{S}_p$.

We use A^+ to denote the Moore–Penrose inverse of matrix A and Sr(A) the spectral radius of square matrix A. A square matrix A is said to be tripotent if $A^3 = A$.

For an n × p matrix Y, Y is written into $Y = [y_1, y_2, \ldots, y_n]'$, $y_i \in \mathfrak{R}^p$, where $\mathfrak{R}^p$ is the p dimensional real space, and vec (Y) denotes n_p dimensional vector $[y'_1, y'_2, \ldots, y'_n]'$. Here the vec operator transforms a matrix into a vector by stacking the rows of the matrix one underneath the other. The Kronecker product $A \otimes B$ of matrices A and B is defined to be $A \otimes B = [a_{ij}B]$. And $(A \otimes B)$ vec $(C) =$ vec (ACB').

The commutation matrix K_{np} is defined by K_{np} vec(Y') = vec(Y), Y ∈ $\mathcal{M}_{n \times p}$. And $\Sigma_{Y'} = K'_{np}\Sigma_Y K_{np}$, see Magnus and Neudecker [10]. The following lemma, see Rao [18, Chapter 1], is useful for our subsequent discussion.

Lemma 2.1: For matrices A, B and C, $AB'B = CB'B$ is equivalent to AB′ = CB′. Similarly, $B'BA = B'BC$ is equivalent to BA = BC. Define

$$G(s, \tilde{s}, \Sigma, W, \Sigma_Y) = \Sigma_Y(W \otimes s)\Sigma_Y(W \otimes \Sigma^+)\Sigma_Y(W \otimes \tilde{s})\Sigma_Y, \quad \text{for } s, \tilde{s} \in \mathcal{S}_p,$$

and

$$\Gamma(s, \tilde{s}, \Sigma, W, L) = L(s \otimes W)L'L(\Sigma^+ \otimes W)L'L(\tilde{s} \otimes W)L', \quad \text{for } s, \tilde{s} \in \mathcal{S}_p.$$

Using the commutation matrix, the properties of the Kronecker product and Lemma 2.1, the following lemma is easily proved.

Lemma 2.2: Let $\Sigma_Y, \in_{np}$, $\Sigma \in \mathcal{N}_p$ and $W \in \mathcal{S}_n$. Then for s, s̃ ∈ $\mathcal{S}_p$,

$$\Sigma_Y[W \otimes (s\Sigma\tilde{s} + \tilde{s}\Sigma s)]\Sigma_Y = G(s, \tilde{s}, \Sigma, W, \Sigma_Y) + G(\tilde{s}, s, \Sigma, W, \Sigma_Y)$$

is equivalent to

$$L[(s\Sigma\tilde{s} + \tilde{s}\Sigma s) \otimes W]L' = \Gamma(s, \tilde{s}, \Sigma, W, L) + \Gamma(s, \tilde{s}, \Sigma, W, L),$$

where $\Sigma_Y = L'L$, $L = [L_1, L_2, \ldots, L_p]$, $q = \text{rank}(\Sigma_Y)$ and $L_i \in \mathcal{M}_{q \times n}$, $i = 1, 2, \ldots, p$.

When we decompose Σ_Y as $\Sigma_Y = L'L$, $L = [L_1, L_2, L_p]$ with $L_i \in \mathcal{M}_{q \times n}$ ($i = 1, 2, \ldots, p$) and $r(\Sigma_Y) \leq q \leq np$, we assume $q = np$ without loss of the generality in our discussion. If $q < np$, we just replace L' by $[L', 0] \in \mathcal{M}_{np \times np}$.

Suppose that $\Sigma_1, \Sigma_2, \ldots, \Sigma_r$ are positive real numbers. Let $B_{ij} = (L_i W L'_j + L_j W L'_i)/2 \sqrt{\Sigma_i \Sigma_j}$, $i, j \leq r$. Assume that $L_i W L'_j \neq 0$, i.e. $B_{ii} \neq 0$, ($i \leq r$) also without loss of generality. For convenience, the following conditions (A1)–(A5) are called A-conditions.

(A1) $\quad L[(t\Lambda\tilde{t} + \tilde{t}\Lambda t) \otimes W]L' = \Gamma(t, \tilde{t}, \Lambda, W, L) + \Gamma(\tilde{t}, t, \Lambda, W, L)$;

(A2) $\quad L(\Lambda^+ \otimes W)L'L(t \otimes W)L' = L(t \otimes W)L'L(\Lambda^+ \otimes W)L'$;

(A3) $\quad \{t : L(t \otimes W)L' = 0\} = \{t : \Lambda t \Lambda = 0\}$;

(A4) $\quad \text{tr}(L(\Lambda^+ \otimes W)L'L(t \otimes W)L') + \text{tr}(L(t \otimes W)L') = 2m_1 \text{tr}(\Lambda t)$

(A5) $\quad \text{tr}(L(\Lambda^+ \otimes W)L'L(t \otimes W)L') - \text{tr}(L(t \otimes W)L') = 2m_2 \text{tr}(\Lambda t)$.

The following conditions (C1)–(C6) are called C-conditions.

(C1) $\quad L_i W L'_j + L_j W L'_i = \mathbf{0}$ for i or $j > r$;

(C2) $\quad B_{ii}^3 = B_{ii}$, $\text{tr}(B_{ii}) = m_1 - m_2$, $\text{tr}(B_{ii}^2) = m_1 + m_2$;

(C3) $\quad B_{ii}B_{jj} = \mathbf{0}$, $i \neq j$;

(C4) $\quad 4B_{ij}^2 = B_{ii}^2 + B_{jj}^2$, $i \neq j$;

(C5) $\quad B_{ii}B_{ij} = B_{ij}B_{jj}$, $i \neq j$;

(C6) $\quad 2(B_{ii} + B_{jj})(B_{ik}B_{jk} + B_{jk}B_{ik}) = B_{ij}$ for all distinct i, j, k.

If $A = X'X$, where $X \sim \mathcal{N}_{m \times p}(, I_m \otimes \Sigma)$ with $\Sigma \in \mathcal{N}_{p'}$, then A is said to have the noncentral Wishart distribution with m degrees of freedom, covariance matrix Σ and no centrality matrix $\lambda = \mu'\mu$. Write $A \sim W_p(m, \Sigma, \lambda)$. When $\mu = 0$, A is said to have the Wishart distribution with m degrees of freedom and covariance matrix Σ, denoted by $A \sim W_p(m, \Sigma)$.

For a symmetric matrix W, Y'WY is called the matrix quadratic form in a normal random matrix Y. $Y{\scriptstyle\rangle}WY \sim W_p(m_1, \Sigma, \lambda_1) - W_p(m_2, \Sigma, \lambda_2)$ means that Y'WY is distributed as a difference of two independent noncentral Wishart random matrices (with a common covariance Σ), implying that $Y{\scriptstyle\rangle}WY$ has the noncentral generalized Laplacianness. Similarly, $Y{\scriptstyle\rangle}WY \sim W_p(m_1, \Sigma) - W_p(m_2, \Sigma)$ means that $Y{\scriptstyle\rangle}$ WY is distributed as a difference of two independent Wishart random matrices (with a common covariance Σ), implying that $Y{\scriptstyle\rangle}$ WY has the generalized Laplacianness. The moment generating function M(s) of $Y{\scriptstyle\rangle}WY$ is defined as $M(s) = E\,(e^{\langle s, Y'WY \rangle})$, $s \in \mathcal{S}_p$. The following lemma is due to Wong et al. [23].

Lemma 2.3: Let $Y \sim \mathcal{N}_{n \times p}(\mu, \Sigma_Y)$ and W_i's be symmetric. Then the joint moment generating function M(s) of $Y'W_iY$'s is given by

$$M(s) = |I_{np} - 2\Sigma^*|^{-1/2}\exp\left\{\langle s, \lambda \rangle + 2\left\langle \mu^*, \Sigma_Y^{1/2}(I_{np} - 2\Sigma^*)^{-1}\Sigma_Y^{1/2}\mu^*\right\rangle\right\},$$

Where $\mathcal{S} = \mathcal{S}_p \times \mathcal{S}_p \times ... \times \mathcal{S}_p$ (l times)

$$s = (s_i) \in \mathcal{S}, \quad \Sigma^* = \Sigma_Y^{1/2}[\textstyle\sum_{i=1}^l (W_i \otimes s_i)]\Sigma_Y^{1/2},$$

$$\mu^* = \textstyle\sum_{i=1}^l vec(W_i\mu s_i),$$

$$\lambda_i = \mu'W_i\mu \in \mathcal{S}_p,$$

$$\lambda = (\lambda_i) \in \mathcal{S}$$

and

$$Sr(\Sigma^*) < 1/2.$$

Let $Y \sim \mathcal{N}_{m \times p}(\mu, I_m \otimes \Sigma)$, then M(s) of $Y{\scriptstyle\rangle}Y$, i.e. the moment generating function of the noncentral Wishart distribution $W_p(m, \Sigma, \lambda)$, is equal to

$$M(s) = |I_p - 2\Sigma_*|^{-m/2}\exp\{\langle s, \lambda \rangle + 2\langle \lambda, s\Sigma^{1/2}(I_p - 2\Sigma_*)^{-1}\Sigma^{1/2}s \rangle\} \quad (2.1)$$

for all $s \in \mathcal{S}_p$ such that $Sr(\Sigma*) < 1/2$ with $\lambda = \mu'\mu$, where $\Sigma* = \Sigma^{1/2} s\Sigma^{1/2}$. And if $Y'WY \sim W_p(m_1, \Sigma, \lambda_1) - W_p(m_2, \Sigma, \lambda_2)$, the moment generating function $M(s)$ of $Y'WY$ can be expressed as

$$M(s) = |I_p - 2\Sigma*|^{-m_1/2}|I_p + 2\Sigma*|^{-m_2/2}\exp\{\langle s, \lambda_1 - \lambda_2 \rangle + 2\Phi_1 + 2\Phi_2\} \qquad (2.2)$$

for all $s \in \mathcal{S}_p$ such that $Sr(\Sigma*) < 1/2$, where $\phi_1 = \langle \lambda_1, s\sum{}^{1/2}(I_p - 2\sum{}_*)^{-1}\sum{}^{1/2}s\rangle$ and $\phi_2 = \langle \lambda_2, s\sum{}^{1/2}(I_p + 2\sum{}_*)^{-1}\sum{}^{1/2}s\rangle$.

We can extend (2.2) so that the case $m_1 = 0$ or $m_2 = 0$ or $\Sigma = 0$ is included. The following result is useful for us to discuss the independence of random matrices, see Hu [6, 7].

Lemma 2.4: Let $Y \sim N_{n\times p}(\mu, \Sigma_Y)$, and W_i's be a family of symmetric matrices in S_n. Then a family of matrix quadratic form $Y'W_iY$'s is independent if and only if for any distinct $i, j \in \{1, 2,...,l\}$ and any $t, \tilde{t}$ in the basic base $\mathcal{B}_{p'}$

1. $\Sigma_Y(W_i \otimes t)\Sigma_Y(W_j \otimes \tilde{t})\Sigma_Y = 0$;

2. $\Sigma_Y(W_i \otimes t)\Sigma_Y(W_j \otimes \tilde{t})vec(\mu) = 0$;

3. $vec(\mu)'(W_i \otimes t)\Sigma_Y(W_j \otimes \tilde{t})vec(\mu) = 0$.

ALGEBRAIC CONDITIONS FOR THE GL OF A MATRIX QUADRATIC FORM

In this section as well as next section, Y is an $n \times p$ multivariate normal random matrix with mean 0 and general covariance Σ_Y

Our investigation begins with the following main theorem. We shall establish a class of sufficient and necessary algebraic conditions to characterize the GL of a matrix quadratic form, i.e. a matrix quadratic form Y'WY being distributed as the difference of two independent Wishart random matrices. The two Wishart distribution have a common covariance Σ.

First let us consider the special case which the common covariance is a diagonal nonnegative definite matrix, written Λ.

Theorem 3.1: Let $Y \sim \mathcal{N}_{n \times p}(0, \Sigma_Y)$ with $\Sigma_Y \in \mathcal{N}_{np}$ and W be symmetric. Then $Y'WY \sim W_p(m_1, \Lambda) - W_p(m_2, \Lambda)$ with $m_1, m_2 \in \{0, 1, 2, ...\}$ and $\Lambda = \text{diag}[\Sigma_1, \Sigma_2, ..., \Sigma_r, 0] \in \mathcal{N}_p$ if and only if there exist positive real numbers $\Sigma_1, \Sigma_2, ..., \Sigma_r$ ($r \leq p$) such that, for any elements $t, \tilde{t}$ in the basic base $\mathcal{B}_{p'}$

$$\Sigma_Y[W \otimes (t\Lambda\tilde{t} + \tilde{t}\Lambda t)]\Sigma_Y = G(t, \tilde{t}, \Lambda, W, \Sigma_Y) + G(\tilde{t}, t, \Lambda, W, \Sigma_Y) \tag{3.1}$$

Such that

$$\Sigma_Y(W \otimes \Lambda^+)\Sigma_Y(W \otimes t)\Sigma_Y = \Sigma_Y(W \otimes t)\Sigma_Y(W \otimes \Lambda^+)\Sigma_Y, \tag{3.2}$$

$$\{t : \Sigma_Y(W \otimes t)\Sigma_Y = 0\} = \{t : \Lambda t \Lambda = 0\}, \tag{3.3}$$

$$\text{tr}(\Sigma_Y(W \otimes \Lambda^+)\Sigma_Y(W \otimes t)) + \text{tr}(\Sigma_Y(W \otimes t)) = 2m_1\text{tr}(\Lambda t) \tag{3.4}$$

and

$$\text{tr}(\Sigma_Y(W \otimes \Lambda^+)\Sigma_Y(W \otimes t)) - \text{tr}(\Sigma_Y(W \otimes t)) = 2m_2\text{tr}(\Lambda t). \tag{3.5}$$

Proof: Exactly as in the proof of Lemma 2.2, we easily derive the equivalence between A-conditions and (3.1)–(3.5). Then by Lemma 7.3, see Appendix, we only show that A-conditions are equivalent to C-conditions.

Suppose that C-conditions hold. We shall show that A-conditions hold. Let $B = \sum_{i=1}^{r} B_{ij}$. Use (ij, i'j') to represent combination $(t, \tilde{t})$ from the basic base $\mathcal{B}$. Then by (C1) we only consider these combinations (ij, i'j'), $1 \leq i \leq j' \leq r$, $1 \leq i' \leq j' \leq r$. Write $\Omega = \{(ij, i'j'): 1 \leq i \leq j \leq r, 1 \leq i' \leq j' \leq r\}$. Divide the index set Ω into the following seven index subsets:

$D_1 = \{(ii, ii) : 1 \leqslant i \leqslant r\};$

$D_2 = \{(ij, ij) : 1 \leqslant i < j \leqslant r\};$

$D_3 = \{(ii, jj) : 1 \leqslant i, j \leqslant r; i \neq j\};$

$D_4 = \{(ii, ij) \cup (ij, ii) : 1 \leqslant i < j \leqslant r\};$

$D_5 = \{(ik, jk) : 1 \leqslant i, j < k \leqslant r; i, j \text{ distinct}\};$

$D_6 = \{(ii, i'j') \cup (i'j', ii) : 1 \leqslant i, i', j' \leqslant r; i, i', j' \text{ distinct}, i' < j'\};$

and

$D_7 = \{(ij, i'j') : 1 \leqslant i < j \leqslant r, 1 \leqslant i' < j' \leqslant r; i, j, i', j' \text{ distinct}\}.$

Note that by (C3), (C4) and Lemma 2.1,

$B_{ij}B_{kk} = \mathbf{0}$ for distinct $i, j, k.$ (3.6)

For (ij, i'j') $\in$ D1, (A1) reduces to $\Sigma_i \Sigma_j (B_{ii} + B_{jj}) = \Sigma^2{}_i B_{ii}BB_{ii}$, which follows from (C2) and (C3).

For (ij, i'j') $\in$ D2, (A1) reduces to $\Sigma_i \Sigma_j (B_{ii} + B_{jj}) = 4\Sigma_i \Sigma_j B_{ij}BB_{ij}$, which is derived from (C5) and (3.6).

For (ij, i'j') $\in$ D3, (A1) reduces to $\Sigma_i \Sigma_j (B_{ii}BB_{jj} + B_{jj}BB_{ii}) = 0$, which is obtained from (C3).

For (ij, i'j') $\in$ D4, (A1) reduces to $2\sqrt{\sigma_i \sigma_j}\sigma_i B_{ij} = 2\sqrt{\sigma_i \sigma_j}\sigma_i(B_{ii}BB_{ij} + B_{ij}BB_{ii})$ which follows from (C5), (C6) and (3.6).

For (ij, i'j') $\in$ D5, (A1) reduces to $2\sqrt{\sigma_i \sigma_j}\sigma_k B_{ij} = 4\sqrt{\sigma_i \sigma_j}\sigma_k(B_{ik}BB_{jk} + B_{jk}BB_{ik})$ which is gotten from (C5), (C6) and (3.6).

For (ij, i'j') $\in$ D6 $\cup$ D7, (A1) reduces to $4\sqrt{\sigma_i \sigma_j \sigma_{i'} \sigma_{j'}} (B_{ij}BB_{i'j'} + B_{i'j'}BB_{ij}) = 0$, which follows from (3.6). So (A1) holds.

For (ij, i'j') $\in \Omega$, (A2) or $BB_{ij} = B_{ij}B$ follows from (3.6) and (C5).

(C1) tells us the fact that $\{t : \Lambda t \Lambda = 0\} = \{t_{ij} \in \mathcal{B}_p : i > r \text{ or } j > r\} \subseteq \{t : L(t \otimes W)L' = 0\}$. For $t_{ij} \in \mathcal{B}_{p'}$, $1 \leq i \leq j \leq r$, $L(t_{ij} \otimes W) L' = 2\sqrt{\sigma_i \sigma_j}B_{ij} \neq 0$ from (C2), (C4) and Lemma 2.1. So $\{t : \Lambda t \Lambda = 0\} = \{t : L(t \otimes W)L' = 0\} = \{t_{ij} \in \mathcal{B}_p : i > r \text{ or } j > r\}$, which proves (A3).

Finally, for $E_{ii} \in \mathcal{B}_{p'}$, $i = 1, 2, ..., r$, with simple calculation, (A4) and (A5) hold by (C2) and (C3). For B_{ij} i, j > r, it is a trivial thing. For B_{ij} i≠ j, i, j ≤r, the right side values of (A4) and (A5) always are zero. We only need to calculate the left side values (LSVs) of both (A4) and (A5). By (A.1) and (A.2) in Lemma 7.2, we have

$$
\begin{aligned}
\text{LSVs} \ &= \ \text{tr}(L(\Lambda^+ \otimes W)L'L(B_{ij} \otimes W)L') \pm \text{tr}(L(B_{ij} \otimes W)L') \\
&= \ 2\sqrt{\sigma_i \sigma_j}[\text{tr}(BB_{ij}) \pm \text{tr}(B_{ij})] = 2\sqrt{\sigma_i \sigma_j}\text{tr}((B_{ii} + B_{jj} \pm I)B_{ij}) \\
&= \ \sqrt{\sigma_i \sigma_j}\text{tr}[(e_{ii} \otimes A_{ii} + e_{jj} \otimes A_{jj} \pm I)(e_{ij} \otimes A_{ij} + e_{ji} \otimes A_{ji})] = 0,
\end{aligned}
$$

which proves (A4) and (A5). Hence A-conditions hold. Conversely, suppose A-conditions hold. We must prove that C-conditions hold. (C1) follows from (A3), i.e. L $(B_{ij} \otimes W)L' = 0$, for i or j > 0. Fixing $(1 \leq i < j \leq r)$ and taking t = $\tilde{t}$ = E_{ii} in (A1) and (A2), we have B

$$
B_{ii} = B_{ii}BB_{ii}, \quad B_{ii}B = BB_{ii} \tag{3.7}
$$

and

$$
\text{tr}(B_{ii}) = m_1 - m_2, \quad \text{tr}(BB_{ii}) = m_1 + m_2. \tag{3.8}
$$

Taking t = E_{ii} and $\tilde{t}$ = E_{jj} in (A1) gives

$$
B_{ii}BB_{jj} + B_{jj}BB_{ii} = 0. \tag{3.9}
$$

By (3.7) and (3.9), we get $||B_{ii}B_{jj} \pm B_{jj}B_{ii}||^2 = 0$, i.e. $B_{ii}B_{jj} = 0$, which proves (C3). Further, we have $B^3 = B$, then $B^3_{ii} = B_{ii}$ and tr $(B^2_{ii}) = \text{tr}(B_{ii}B_{ii}BB_{ii}) = \text{tr}(B^3_{ii}B) = \text{tr}(B_{ii}B) = m_1 + m_2$, which, with (3.8), proves (C2).

Taking t = E_{ii} and $\tilde{t}$ = B_{ij} in (A1) gives $B_{ij} = B_{ii}BB_{ij} + B_{ij}BB_{ii}$. Taking t = B_{ij} and $\tilde{t}$ = E_{jj} in (A1) and (A2) gives

$$
B_{ij} = B_{ij}BB_{jj} + B_{jj}BB_{ij}, \quad BB_{ij} = B_{ij}B. \tag{3.10}
$$

So $B_{ii}B_{ij} = B_{ii}B_{ij}BB_{jj}$ and $B_{ij}B_{jj} = B_{ii}BB_{ij}B_{jj}$, which proves (C5). Taking t = $\tilde{t}$ = B_{ij} in (A1) gives

$$
4B_{ij}BB_{ij} = B_{ii} + B_{jj}. \tag{3.11}
$$

From (C3), (C5) and (3.10)–(3.11), we obtain $4B^2_{ij} = B^2_{ii} + B^2_{jj}$, which proves (C4). From (3.10), (C3) and (C5), we obtain, for distinct i, j, k,

$$B_{ij}B_{kk} = 0. \tag{3.12}$$

Taking $t = E_{ik}$ and $\tilde{t} = E_{jk}$ for distinct i, j, k in (A1) gives

$$B_{ij} = 2B_{ik}BB_{jk} + 2B_{jk}BB_{ik}. \tag{3.13}$$

So from (3.13), (3.12) and (C5), we get $B_{ij} = 2(B_{ii} + B_{jj})(B_{ik}B_{jk} + 2B_{jk}B_{ik})$, which proves (C6). Thus the desired result is proved.

In Theorem 3.1, condition (3.1) reveals the most important inherent property for the GL of a matrix quadratic form. Condition (3.2) tells us that matrix $\sum_Y^{1/2}(W \otimes \Lambda+)\Sigma^{1/2}_Y$ is commutative with any matrix $\sum_Y^{1/2}$ $(W \otimes t)\sum_Y^{1/2}$ for any t in the basic base $\mathcal{B}_p$. Condition (3.3) says that two different linear transformations $\Sigma_Y(W \otimes t)\Sigma_Y$ and $\Lambda t\Lambda$ have the same kernel or null space. Conditions (3.4) and (3.5), respectively, determines the degrees m_1, m_2 of freedom of two Wishart random matrices. Next we shall extend Theorem 3.1 to the general case which the common covariance is a general nonnegative definite matrix Σ. A set of the corresponding sufficient and necessary algebraic conditions is summarized in the following theorem.

Theorem 3.2: Suppose that $Y \sim \mathcal{N}_{n \times p}(0, \Sigma_Y)$ with $\Sigma_Y \in \mathcal{N}_{np}$ and W is symmetric. Then $YW_Y \sim W_p(m_1, \Sigma) - Wp(m_2, \Sigma)$ with $m_1, m_2 \in \{0, 1, 2, ...\}$ and $\Sigma \in N_p$ if and only if there exists some $\Sigma \in \mathcal{N}_p$ such that, for any elements h, $\hat{h}$ in the similar base $\mathcal{H}_p$ associated with Σ,

$$\Sigma_Y[W \otimes (h\Sigma\tilde{h} + \tilde{h}\Sigma h)]\Sigma_Y = G(h, \tilde{h}, \Sigma, W, \Sigma_Y) + G(\tilde{h}, h, \Sigma, W, \Sigma_Y) \tag{3.14}$$

Such that

$$\Sigma_Y(W \otimes \Sigma^+)\Sigma_Y(W \otimes h)\Sigma_Y = \Sigma_Y(W \otimes h)\Sigma_Y(W \otimes \Sigma^+)\Sigma_Y, \tag{3.15}$$

$$\{h : \Sigma_Y(W \otimes h)\Sigma_Y = 0\} = \{h : \Sigma h\Sigma = 0\}, \tag{3.16}$$

$$tr(\Sigma_Y(W \otimes \Sigma^+)\Sigma_Y(W \otimes \mathbf{h})) + tr(\Sigma_Y(W \otimes \mathbf{h})) = 2m_1 tr(\Sigma \mathbf{h}) \qquad (3.17)$$

And

$$tr(\Sigma_Y(W \otimes \Sigma^+)\Sigma_Y(W \otimes \mathbf{h})) - tr(\Sigma_Y(W \otimes \mathbf{h})) = 2m_2 tr(\Sigma \mathbf{h}). \qquad (3.18)$$

Proof: If $\Sigma \in \mathcal{N}_p$, there is an orthogonal matrix H of order p such that

$$H'\Sigma H = diag[\sigma_1, \sigma_2, \ldots, \sigma_r, \mathbf{0}] \equiv \Lambda, \quad r = r(\Sigma), \quad \sigma_i > 0, \quad i = 1, 2, \ldots, r \qquad (3.19)$$

and $YH \sim \mathcal{N}_{n \times p}(0, \Sigma_{YH})$ where $\Sigma_{YH} = (I \otimes H')\Sigma_Y(I \otimes H)$.

Assume that there exists $\Sigma \in \mathcal{N}_p$ such that (3.14)–(3.18) hold. Let t = H'hH, then function t= H'hH' is a 1–1 map from the similar base $\mathcal{H}_p$ associated with Σ onto the basic base $\mathcal{B}_p$. By replacing h, h̃ and Σ, respectively, with HtH', Ht̃H' and HΛH' in (3.14)–(3.18), with necessary Kronecker product operations, (3.14)–(3.18) are equivalent to the following equations, for t, t̃ $\in \mathcal{B}_p$,

$$\Sigma_{YH}\left[W \otimes (t\Lambda \tilde{t} + \tilde{t}\Lambda)\right]\Sigma_{YH} = G(t, \tilde{t}, \Lambda, W_i, \Sigma_{YH}) + G(\tilde{t}, t, \Lambda, W_i, \Sigma_{YH}),$$

$$\Sigma_{YH}(W \otimes \Lambda^+)\Sigma_{YH}(W \otimes t)\Sigma_{YH} = \Sigma_{YH}(W \otimes t)\Sigma_{YH}(W \otimes \Lambda^+)\Sigma_{YH},$$

$$\{t : \Sigma_{YH}(W \otimes t)\Sigma_{YH} = 0\} = \{t : \Lambda t\Lambda = 0\},$$

$$tr(\Sigma_{YH}(W \otimes \Lambda^+)\Sigma_{YH}(W \otimes t)) + tr(\Sigma_{YH}(W \otimes t)) = 2m_1 tr(\Lambda t)$$

and

$$tr(\Sigma_{YH}(W \otimes \Lambda^+)\Sigma_{YH}(W \otimes t)) - tr(\Sigma_{YH}(W \otimes t)) = 2m_2 tr(\Lambda t),$$

where Λ is determined by (3.19).

By Theorem 3.1, H'Y'WYH $\sim W_p(m_1, \Lambda) - W_p(m_2, \Lambda)$. Hence Y'WY $\sim W_p(m_1, \Sigma) - W_p(m_2, \Sigma)$ by Theorem 3.2.4 of Moorhead (1982). The converse can be shown by following the above steps backwards.

Conditions (3.14)–(3.18) are same as the conditions (3.1)–(3.5) except for using the similar base $\mathcal{H}_p$ associated with Σ to replace the basic base $\mathcal{B}_p$ of $\mathcal{S}_p$.

Whenever $Y'WY \sim W_p(m_1, \Sigma) - W_p(m_2, \Sigma)$, then the degrees m_1, m_2 of freedom are determined by

$$m_1 = [\text{tr}(\Sigma_Y(W \otimes \Sigma^+))^2 + \text{tr}(\Sigma_Y(W \otimes \Sigma^+))]/2r(\Sigma) \text{ and}$$
$$m_2 = [\text{tr}(\Sigma_Y(W \otimes \Sigma^+))^2 - \text{tr}(\Sigma_Y(W \otimes \Sigma^+))]/2r(\Sigma), \tag{3.20}$$

where $r(A)$ denotes the rank of matrix A.

In Theorem 3.2, if y is an $n \times 1$ random normal vector with mean vector 0 and covariance C of order n, the conditions (3.14)–(3.18) reduce to the familiar algebraic conditions.

Corollary 3.3: Let $y \sim \mathcal{N}_n(0, C)$ with $C \in \mathcal{N}_n$ and W be a symmetric matrix of order n. Then $y'Wy \sim \chi^2(m_1) - \chi^2(m_2)$, a difference of two independent chi-square random variables, with m_1, $m_2 \in \{0, 1, 2,...\}$ if and only if

1. $CWC = CWCWCWC \neq 0$;
   and

2. $\text{tr}(CW)^2 + \text{tr}(CW) = 2m_1,\ \text{tr}(CW)^2 - \text{tr}(CW) = 2m_2.$

If $C = I$ in Corollary 3.3, statement (1) of Corollary 3.3 reduces to the well-known trip tent condition, $W^3 = W$, which is the necessary and sufficient condition to a quadratic form $y'Wy$ being distributed as a difference of two independent chi-squared random variables, see Luther [9] and Gray bill [4].

ALGEBRAIC CONDITIONS FOR THE GL AND INDEPENDENCE OF A FAMILY OF MATRIX QUADRATIC FORMS

Based on Theorem 3.1 and (1) of Lemma 2.4, we shall establish the following result on the GL and independence of a family of matrix quadratic forms. In other words, we shall provide an extension of Cochran's theorem concerning the GL and independence of a family of matrix quadratic forms. Similar to the discussion in Section 3, first let us consider the simplest case where the common covariance Λ of independent Wishart random matrices is diagonal.

Theorem 4.1: Suppose that $Y \sim \mathcal{N}_{n \times p}(0, \Sigma_Y)$ with $\Sigma_Y \in {}_{np}$ and W_i's are symmetric matrices of order n. Then $Y' W_1 Y, Y' W_2 Y, \ldots, Y' W_l Y$ are independent and, for any $i \in \{1, 2, \ldots, l\}$, $Y' W_i Y \sim W_p(m_{1i}, \Lambda) - W_p(m_{2i}, \Lambda)$ with $m_{1i}, m_{2i} \in \{0, 1, 2, \ldots\}$ and $\Lambda = \text{diag}[\Sigma_1, \Sigma_2, \ldots, \Sigma_r, 0] \in \mathcal{N}_p$ if and only if there exist positive real numbers $\Sigma_1, \Sigma_2, \ldots, \Sigma_r$ ($r \leq p$) such that, for any distinct $i, j \in \{1, 2, \ldots, l\}$ and any $t, \tilde{t}$ in the basic base $\mathcal{B}_{p'}$

a) $\Sigma_Y[W_i \otimes (t \Lambda \tilde{t} + \tilde{t} \Lambda t)] \Sigma_Y = G(t, \tilde{t}, \Lambda, W_i, \Sigma_Y) + G(\tilde{t}, t, \Lambda, W_i, \Sigma_Y)$;

b) $\Sigma_Y(W_i \otimes \Lambda^+) \Sigma_Y(W_i \otimes t) \Sigma_Y = \Sigma_Y(W_i \otimes t) \Sigma_Y(W_i \otimes \Lambda^+) \Sigma_Y$;

c) $\{t : \Sigma_Y(W_i \otimes t) \Sigma_Y = 0\} = \{t : \Lambda t \Lambda = 0\}$;

d) $\text{tr}(\Sigma_Y(W_i \otimes \Lambda^+) \Sigma_Y(W_i \otimes t)) + \text{tr}(\Sigma_Y(W_i \otimes t)) = 2m_{1i}\text{tr}(\Lambda t)$,
$\text{tr}(\Sigma_Y(W_i \otimes \Lambda^+) \Sigma_Y(W_i \otimes t)) - \text{tr}(\Sigma_Y(W_i \otimes t)) = 2m_{2i}\text{tr}(\Lambda t)$;

e) $\Sigma_Y(W_i \otimes \Lambda^+) \Sigma_Y(W_j \otimes \Lambda^+) \Sigma_Y = 0$.

Proof: Let $\{Y' W_i Y\}_{i=1}^l$ be an independent family of random matrices and $Y' W_i Y \sim W_p(m_{1i}, \Lambda) - W_p(m_{2i}, \Lambda)$ with $m_{1i}, m_{2i} \in \{0, 1, 2, \ldots\}$ and $\Lambda = \text{diag}[\Sigma_1, \Sigma_2, \ldots, \Sigma_r, 0] \in \mathcal{N}_{p'}$ $i = 1, 2, \ldots, l$. Then (a)–(e) follow from Theorem 3.1 and (1) of Lemma 2.4.

Conversely, suppose that there exist positive real numbers $\Sigma_1, \Sigma_2,\ldots,$ Σ_r ($r \leq p$) such that (a)–(e) hold. For each i, from Theorem 3.1, $Y' W_i Y \sim W_p(m_{1i}, \Lambda) - W_p(m_{2i}, \Lambda)$ with $m_{1i}, m_{2i} \in \{0, 1, 2, \ldots\}$ and $\Lambda = \text{diag}[\Sigma_1, \Sigma_2,\ldots, \Sigma_r, 0] \in {}_p$. To prove the independence of a family of matrix quadratic forms, by Lemma 2.4, it suffices to show (a) of Lemma 2.4, or equivalently,

$$\Sigma_Y(W_i \otimes s_i) \Sigma_Y(W_j \otimes s_j) \Sigma_Y = 0, \quad \text{for } s_i, s_j \in \mathcal{S}_p, \qquad (4.1)$$

from conditions (a)–(e).

Exactly as in the proof of Lemma 2.2, (4.1) is equivalent to

$$L(\mathbf{s}_i \otimes W_i)L'L(\mathbf{s}_j \otimes W_j)L' = \mathbf{0}, \quad \text{where } L'L = \Sigma_{Y'}, \ \mathbf{s}_i, \mathbf{s}_j \in \mathcal{S}_p \tag{4.2}$$

and condition (e) amounts to

$$L(\Lambda^+ \otimes W_i)L'L(\Lambda^+ \otimes W_j)L' = \mathbf{0}. \tag{4.3}$$

Then we only need to obtain (4.2) from statements (a)–(e). For matrix s_i in set $\mathcal{S}_p$, s_i can be written as

$$s_i = \begin{bmatrix} \mathbf{a} & * \\ * & * \end{bmatrix}_{p \times p} \quad \text{where } \mathbf{a} \in \mathcal{S}_r.$$

Write

$$s_i^* = \begin{bmatrix} \mathbf{a} & \mathbf{0} \\ \mathbf{0} & \mathbf{0} \end{bmatrix}_{p \times p} \quad \text{where } \mathbf{a} \in \mathcal{S}_r.$$

By (c), for any s_i, $s_j \in \mathcal{S}_p$,

$$L(\mathbf{s}_i \otimes W_i)L'L(\mathbf{s}_j \otimes W_j)L' = L(\mathbf{s}_i^* \otimes W_i)L'L(\mathbf{s}_j^* \otimes W_j)L'. \tag{4.4}$$

Since, by (a) and (b),

$$\begin{aligned} 2L(\mathbf{s}_i^* \otimes W_i)L' &= L[(\Lambda^+ \Lambda \mathbf{s}_i^* + \mathbf{s}_i^* \Lambda \Lambda^+) \otimes W_i]L' \\ &= [L(\Lambda^+ \otimes W_i)L']^2 L(\mathbf{s}_i^* \otimes W_i)L' + L(\mathbf{s}_i^* \otimes W_i)L'[L(\Lambda^+ \otimes W_i)L']^2 \\ &= 2L(\mathbf{s}_i^* \otimes W_i)L'[L(\Lambda^+ \otimes W_i)L']^2, \end{aligned} \tag{4.5}$$

similarly,

$$L(\mathbf{s}_j^* \otimes W_j)L' = [L(\Lambda^+ \otimes W_j)L']^2 L(\mathbf{s}_j^* \otimes W_j)L', \tag{4.6}$$

we obtain (4.2) from (4.3)–(4.6). So, we have completed the proof.

In Theorem 4.1, condition (e) tells us that one equation can be used to reveal the independence of a set of matrix quadratic forms if these matrix quadratic forms have the GL. Next, we will consider the general case where the common covariance Σ of these Wishart random matrices is nonnegative definite. Exactly as in the proof of Theorem 3.2, we can easily derive the following theorem, an extension of Cochran's theorem concerning the GL and independence of a set of matrix quadratic forms, from Theorem 4.1 and (1) of Lemma 2.4 with an appropriate modification by replacing $\mathcal{H}_p$ with $\mathcal{B}_p$. See Hu [6, Chapter 4].

Theorem 4.2: Suppose that $Y \sim \mathcal{N}_{n \times p}(0, \Sigma_Y)$ with $\Sigma_Y \in \mathcal{N}_{np}$ and W_i's are symmetric matrix of order n. Then $Y' W_1 Y, Y' W_2 Y, \ldots, Y W_1 Y$ are independent and, for any $i \in \{1, 2, \ldots, l\}$, $Y W_i Y \sim W_p(m_{1i}, \Sigma) - W_p(m_{2i}, \Sigma)$ with $m_{1i}, m_{2i} \in \{0, 1, 2, \ldots\}$ and $\Sigma \in \mathcal{N}_p$ if and only if there exists some $\Sigma \in \mathcal{N}_p$ such that, for any distinct $i, j \in \{1, 2, \ldots, l\}$ and any $h, \tilde{h}$ in the similar base $\mathcal{H}_p$ associated with Σ,

a) $\Sigma_Y[W_i \otimes (h\Sigma\tilde{h} + \tilde{h}\Sigma h)]\Sigma_Y = G(h, \tilde{h}, \Sigma, W_i, \Sigma_Y) + G(\tilde{h}, h, \Sigma, W_i, \Sigma_Y);$

b) $\Sigma_Y(W_i \otimes \Sigma^+)\Sigma(W_i \otimes h)\Sigma_Y = \Sigma_Y(W_i \otimes h)\Sigma_Y(W_i \otimes \Sigma^+)\Sigma_Y;$

c) $\{h : \Sigma_Y(W_i \otimes h)\Sigma_Y = 0\} = \{h : \Sigma h \Sigma = 0\};$

d) $\mathrm{tr}(\Sigma_Y(W_i \otimes \Sigma^+)\Sigma_Y(W_i \otimes h)) + \mathrm{tr}(\Sigma_Y(W_i \otimes h)) = 2m_{1i}\mathrm{tr}(\Sigma h),$
$\mathrm{tr}(\Sigma_Y(W_i \otimes \Sigma^+)\Sigma_Y(W_i \otimes h)) - \mathrm{tr}(\Sigma_Y(W_i \otimes h)) = 2m_{2i}\mathrm{tr}(\Sigma h);$ and

e) $\Sigma_Y(W_i \otimes \Sigma^+)\Sigma_Y(W_j \otimes \Sigma^+)\Sigma_Y = 0.$

In Theorem 4.2, if we replace covariance Σ_Y of Y with the sum of special Kronecker products, we have the following corollary, an application of Theorem 4.2 on a special case. See Hu [6, Chapter 4].

Corollary 4.3: Let $Y \sim \mathcal{N}_{n \times p}(0, \Sigma_Y)$ with $\Sigma_Y = \sum_{a=1}^{r} A_a \otimes E_{aa}$, $r p, A_a \in \mathcal{N}_{n'}$ and W_i's are symmetric matrices of order n. Then $Y'W_1 Y, Y'W_2 Y, \ldots, Y'W_l Y$ are independent and, for any $i \in \{1, 2, \ldots, l\}$, $Y' W_i Y \sim W_p(m_{1i}, \Sigma) - W_p(m_{2i}, \Sigma)$ with $m_{1i}, m_{2i} \in \{0, 1, 2, \ldots\}$ and $\Sigma = \sum_{b=1}^{r} \sigma_b E_{bb}$ if and only if there exist positive real numbers $\Sigma_1, \Sigma_2, \ldots, \Sigma_r$ ($r \leq p$) such that, for all $a, b, c \in \{1, 2, \ldots, r\}$ and any distinct $i, j \in \{1, 2, \ldots, l\}$,

1) $A_a W_i A_c W_i A_c W_i A_b = \sigma_c^2 A_a W_i A_b \neq 0;$

2) $\sigma_b A_a W_i A_a W_i A_b = \sigma_a A_a W_i A_b W_i A_b;$

3) $A_a W_i A_a W_j A_a = 0;$

4) $\mathrm{tr}(A_a W_i)^2/\sigma_a^2 + \mathrm{tr}(A_a W_i)/\sigma_a^2 = 2m_{1i};$

5) $\text{tr}(A_a W_i)^2/\sigma_a^2 - \text{tr}(A_a W_i)/\sigma_a^2 = 2m_{2i}$.

CONDITIONS FOR THE NONCENTRAL GL OF A MATRIX QUADRATIC FORM

In this section and next section, Y is an n × p multivariate normal random matrix with nonzero mean μ and general covariance Σ_Y.

We shall use the moment generating function M(s) of Y'WY to extend Theorem 3.2 to the case of Y having nonzero mean. The following theorem summarizes a set of sufficient and necessary algebraic conditions for the noncentral GL of a matrix quadratic form.

Theorem 5.1: Suppose $Y \sim \mathcal{N}_{n\times p}(\mu, \Sigma_Y)$ with $\Sigma_Y \in \mathcal{N}_{np}$ and W is a symmetric matrix. Then $Y'WY \sim W_p(m_1, \Sigma, \lambda_1) - W_p(m_2, \Sigma, \lambda_2)$ with m_1, $m_2 \in \{0, 1, 2, ...\}$, $\Sigma \in \mathcal{N}_p$ and $\lambda_1, \lambda_2 \in \mathcal{M}_{p\times p}$ if and only if there exists some $\Sigma \in \mathcal{N}_p$ such that, in addition to (3.14)–(3.18), for any s in a neighborhood $\mathcal{N}_0$ of 0 in $\mathcal{S}_p$ and k = 1, 2, ...

$$\text{tr}((\lambda_1 + \lambda_2)s(\Sigma s)^{2k-1}) = \text{tr}(\Delta(W \otimes s)[\Sigma_Y(W \otimes s)]^{2k-1}) \qquad (5.1)$$

and

$$\text{tr}((\lambda_1 - \lambda_2)s(\Sigma s)^{2k}) = \text{tr}(\Delta(W \otimes s)[\Sigma_Y(W \otimes s)]^{2k}) \qquad (5.2)$$

With

$$\lambda_1 - \lambda_2 = \mu'W\mu \qquad (5.3)$$

where $\Delta = \text{vec}(\mu)\,\text{vec}(\mu)'$.

Proof: By Lemma 2.3, the moment generating function M(s) of Y'WY is expressed as

$$M(s) = \left|I - 2\Sigma_Y^{1/2}(W \otimes s)\Sigma_Y^{1/2}\right|^{-1/2} \exp\{\langle s, \mu'W\mu\rangle + 2\Phi_0\} \qquad (5.4)$$

where

$$\Phi_0 = \langle \Delta, \ (W \otimes s)\Sigma_Y^{1/2}[I - 2\Sigma_Y^{1/2}(W \otimes s)\Sigma_Y^{1/2}]^{-1}\Sigma_Y^{1/2}(W \otimes s)\rangle$$

and the spectral radius of square matrix $\sum_Y^{1/2}(W \otimes s)\sum_Y^{1/2}$ is less than 1/2. So, $Y'WY \sim W_p(m_1, \Sigma, {}_1) - W_p(m_2, \Sigma, {}_2)$ is equivalent to $Y'WY = D_1 - D_2$, where D_1, D_2 are independent and $D_1 \sim W_p(m_1, \Sigma, {}_1)$, $D_2 \sim W_p(m_2, \Sigma, {}_2)$. By (2.1), $M_1(s)$ of D_1 and $M_2(s)$ of $-D_2$ are given, respectively, by

$$M_1(s) = |I - 2\Sigma_*|^{-m_1/2}\exp\{\langle s, \lambda_1\rangle + 2\Phi_1\} \tag{5.5}$$

and

$$M_2(s) = |I + 2\Sigma_*|^{-m_2/2}\exp\{\langle -s, \lambda_2\rangle + 2\Phi_2\}, \tag{5.6}$$

Where

$$\Sigma_* = \Sigma^{1/2}s\Sigma^{1/2}$$

for $s \in \mathcal{S}_p$ such that $\mathrm{Sr}(\Sigma_*) < 1/2$ and Φ_i's are defined in (2.2). The independence of D_1 and D_2 and (5.4)–(5.6) imply that there exists a neighborhood $\mathcal{N}_0$ of 0 in $\mathcal{S}_p$ such that $M(s) = M_1(s) M_2(s)$ for $s \in \mathcal{N}_0$. Using (5.4)–(5.6) and comparing the same items in both sides of $M(s) = M_1(s) M_2(s)$, we obtain the following conditions:

I. $|I - 2\Sigma_Y^{1/2}(W \otimes s)\Sigma_Y^{1/2}|^{-1/2} = |I - 2\Sigma_*|^{-m_1/2}|I + 2\Sigma_*|^{-m_2/2}$;

II. for any $s \in \mathcal{N}_0$, $\Phi_0 = \Phi_1 + \Phi_2$; and

III. $\lambda_1 - \lambda_2 = \mu'W\mu$,

which proves (5.3) as required.

By Lemma 7.1, the condition (i) is equivalent to $(Y -\mu)\rangle W(Y -\mu) \sim_p (m_1, \Sigma) - W_p (m_2, \Sigma)$. Thus (3.14)–(3.18) follow from Theorem 3.2. For any symmetric matrix $s \in \mathcal{N}_0$, we have

$$\Phi_1 = \mathrm{tr}\left(\lambda_1[s\Sigma s + 2s(\Sigma s)^2 + 2^2 s(\Sigma s)^3 + \cdots]\right), \tag{5.7}$$

$$\Phi_2 = \mathrm{tr}\left(\lambda_2[s\Sigma s - 2s(\Sigma s)^2 + 2^2 s(\Sigma s)^3 - \cdots]\right) \tag{5.8}$$

and

$$\Phi_0 = \mathrm{tr}\left(\Delta[(W \otimes s)\Upsilon + 2(W \otimes s)\Upsilon^2 + 2^2(W \otimes s)\Upsilon^3 + \cdots]\right), \qquad (5.9)$$

where $= \Sigma_Y (W \otimes s)$. Putting (5.7)–(5.9) into the equation $_0 = _1 + _2$ and then comparing its both sides, with the arbitrariness of s close to 0, we obtain (5.1) and (5.2). Thus we have completed the proof of the desired result.

The nonzero mean of Y results in conditions (5.1) and (5.2). In fact, we have obtained the following relation between Y'WY and $(Y - \mu)' W(Y - \mu)$ in the proof of Theorem 5.1.

Corollary 5.2: Let $Y \sim \mathcal{N}_{n \times p}(, \Sigma_Y)$ with $\Sigma_Y \in \mathcal{N}_{np}$ and W be symmetric. Then $Y' WY \sim W_p(m_1, \Sigma, _1) - W_p(m_2, \Sigma, _2)$ with $m_1, m_2 \in \{0, 1, 2,...\}$, $\Sigma \in \mathcal{N}_p$ and $1, 2 \in \mathcal{M}_{p \times p}$ if and only if there exists some $\Sigma \in \mathcal{N}_p$ such that,

a. $(Y - \mu)' W (Y - \mu) \sim W_p(m_1, \Sigma) - W_p(m_2, \Sigma)$; and

b. for any s in a neighborhood $\mathcal{N}_0$ of $0 \in \mathcal{S}_p$ and $k = 1, 2, \ldots$,

$$\mathrm{tr}\left((\lambda_1 + \lambda_2)s(\Sigma s)^{2k-1}\right) = \mathrm{tr}\left(\Delta(W \otimes s)[\Sigma_Y(W \otimes s)]^{2k-1}\right)$$
$$\mathrm{tr}\left((\lambda_1 - \lambda_2)s(\Sigma s)^{2k}\right) = \mathrm{tr}\left(\Delta(W \otimes s)[\Sigma_Y(W \otimes s)]^{2k}\right)$$

with

$$\lambda_1 - \lambda_2 = \mu'W\mu.$$

CONDITIONS FOR THE NONCENTRAL GL AND INDEPENDENCE OF A FAMILY OF MATRIX QUADRATIC FORMS

In this section, we shall use the moment generating function M(s) of Y'WY to extend Theorem 4.2 to the nonzero mean case of Y. Based on Theorem 5.1, we obtain the following extension of Cochran's theorem concerning the noncentral GL and independence of a family of matrix quadratic forms by putting Theorem 4.2, Theorem 5.1 and Lemma 2.4 together with an appropriate modification.

Theorem 6.1: Let $Y \sim \mathcal{N}_{n \times p}(, \Sigma_Y)$ with $\Sigma_Y \in \mathcal{N}_{np}$ and W_i's be symmetric matrices of order n. Then $Y'W_iY \sim W_p(m_{1i}, \Sigma, \lambda_{1i}) - \mathcal{M}_p(m_{2i}, \Sigma, \lambda_{2i})$ with $m_{1i}, m_{2i} \in \{0, 1, 2, ...\}$, $\Sigma \in \mathcal{N}_p$ and $\lambda_{1i}, \lambda_{2i} \in \mathcal{M}_{p \times p}$ if and only if there exists some $\Sigma \in \mathcal{N}_p$ such that, in addition to conditions (a)–(e) of Theorem 4.2, the following statements (f) and (g) also hold.

f   For any distinct i, j $\in \{1, 2,..., l\}$ and $t, \hat{t} \in \mathcal{B}_{p'}$

$$\Sigma_Y(W_i \otimes t)\Sigma_Y(W_j \otimes \tilde{t})\text{vec}(\mu) = 0$$

and

$$\text{vec}(\mu)'(W_i \otimes t)\Sigma_Y(W_j \otimes \tilde{t})\text{vec}(\mu) = 0;$$

and

g   for any s in a neighborhood $\mathcal{N}_0$ of 0 in $\mathcal{S}_{p'}$, i = 1, 2, ... , l and k = 1, 2, ... ,

$$\text{tr}\left((\lambda_{1i} + \lambda_{2i})s(\Sigma s)^{2k-1}\right) = \text{tr}\left(\Delta(W_i \otimes s)[\Sigma_Y(W_i \otimes s)]^{2k-1}\right)$$

and

$$\text{tr}\left((\lambda_{1i} - \lambda_{2i})s(\Sigma s)^{2k}\right) = \text{tr}\left(\Delta(W_i \otimes s)[\Sigma_Y(W_i \otimes s)]^{2k}\right)$$

with

$$\lambda_{1i} - \lambda_{2i} = \mu'W_i\mu.$$

Finally, let us look at a special case $\Sigma_Y = A \otimes \Sigma$ of Theorem 6.1 investigated by Tan [21]

Corollary 6.2: In Theorem 6.1, suppose that $\Sigma_Y = A \otimes \Sigma$ for some $A \in \mathcal{N}_n$ and $\Sigma \in \mathcal{N}_p$. Then $Y'W_1Y, Y'W_2Y, ..., Y'W_lY$ are independent and, for each i, $YW_iY \sim W_p(m_{1i}, \Sigma, \lambda_{1i}) - Wp(m_{2i}, \Sigma, \lambda_{2i})$ with $m_{1i}, m_{2i} \in \{0, 1, 2, ...\}$, $\Sigma \in \mathcal{N}_p$ and $\lambda_{1i}, \lambda_{2i} \in \mathcal{M}_{p \times p}$ if and only if for any distinct $i, j \in \{1, 2, ..., l\}$,

1. $AW_iAW_iAW_iA = AW_iA \neq 0$;

2. $\mathrm{tr}(AW_i)^2 + \mathrm{tr}(AW_i) = 2m_{1i}, \mathrm{tr}(AW_i)^2 - \mathrm{tr}(AW_i) = 2m_{2i}$;

3. $\lambda_{1i} + \lambda_{2i} = \mu'W_iAW_i\mu = \mu'W_iAW_iAW_iAW_i\mu, \lambda_{1i} - \lambda_{2i} = \mu'W_i\mu = \mu'W_iAW_iAW_i\mu$;

4. $AW_iAW_jA = 0$;

5. $AW_iAW_j\mu = 0$; and

6. $\mu'W_iAW_j\mu = 0$.

CONCLUDING REMARKS

In this article, we obtain a set of the sufficient and necessary conditions under which a matrix quadratic form Y'WY with a symmetric W and a general Σ_Y is distributed as a difference of two independent (noncentral) Wishart random matrices, namely, having the (noncentral) GL. Based on the results, we then establish some general extensions of Cochran's theorem concerning the (noncentral) GL and independence of a set of matrix quadratic forms. It should be noted that it is challenging research for us to obtain a set of sufficient and necessary conditions under which a matrix quadratic form Y'WY with symmetric matrix W and a general covariance Σ_Y of Y is distributed as a linear combination of independent Wishart random matrices.

ACKNOWLEDGMENTS

The author would like to appreciate Professor C.S. Wong for his guidance and encouragement in the earlier stage of the article. The author are also deeply grateful to the anonymous referee and Editors-in-chief Richard A. Brualdi for their valuable remarks.

APPENDIX

Lemma 7.1: Let

$$Y \sim \mathcal{N}_{n \times p}(\mathbf{0}, \Sigma_Y)$$

and

$$\Sigma \in \mathcal{N}_p.$$

Then the following statements are equivalent.

a) $Y'WY \sim W_p(m_1, \Sigma) - W_p(m_2, \Sigma)$;

b) for any $s \in \mathcal{S}_{p'}$

$$\left| I_{np} - 2\Sigma_Y^{1/2}(W \otimes s)\Sigma_Y^{1/2} \right| = \left| I_p - 2\Sigma^{1/2}s\Sigma^{1/2} \right|^{m_1} \left| I_p + 2\Sigma^{1/2}s\Sigma^{1/2} \right|^{m_2};$$

c) $\Sigma_Y^{1/2}(W \otimes s)\Sigma_Y^{1/2}$

and
$\mathrm{diag}[I_{m_1} \otimes \Sigma^{1/2}s\Sigma^{1/2}, -I_{m_2} \otimes \Sigma^{1/2}s\Sigma^{1/2}, \mathbf{0}]$ have the same characteristic polynomial for all $s \in \mathcal{S}_{p'}$ and

d) for any positive integer k and any $s \in \mathcal{S}_p$

$$\mathrm{tr}(\Sigma_Y(W \otimes s))^k = \left(m_1 + (-1)^k m_2 \right) \mathrm{tr}(\Sigma s)^k.$$

Proof: Let

$$\Sigma_* = \Sigma^{1/2}s\Sigma^{1/2}$$

From Lemma 2.3 and (2.2) with $\lambda_1 = \lambda_2 = 0$, (a) is equivalent to (a') for
any $s \in \mathcal{S}_p$ such that $S_r(\Sigma_*) < 1/2$ and

$$Sr(\Sigma_Y^{1/2}(W \otimes s)\Sigma_Y^{1/2}) < 1/2,$$

$$\left| I_{np} - 2\Sigma_Y^{1/2}(W \otimes s)\Sigma_Y^{1/2} \right|^{-1/2} = |I_p - 2\Sigma_*|^{-m_1/2}|I_p + 2\Sigma_*|^{-m_2/2}.$$

It's obvious that (a') follows from (b). And (b) is obtained from (a') by analytic continuation. Thus (a) and (b) are equivalent.

Replace s with s/2λ (nonzero $\lambda \in \mathfrak{R}$) in (b) and multiplying both sides of (b) by λ^{np}. Then (b) amounts to

$$\left| \lambda I_{np} - \Sigma_Y^{1/2}(W \otimes s)\Sigma_Y^{1/2} \right| = |\lambda I_p - \Sigma_*|^{m_1}|\lambda I_p + \Sigma_*|^{m_2}|\lambda I_p - 0|^{(n-m_1-m_2)}.$$

So (c) is equivalent to (b). (c) Amounts to that matrix $\Sigma_Y^{1/2}(W \otimes s)\Sigma_Y^{1/2}$ and diagonal matrix

$$\mathrm{diag}[I_{m_1} \otimes \Sigma^{1/2}s\Sigma^{1/2}, -I_{m_2} \otimes \Sigma^{1/2}s\Sigma^{1/2}, 0]$$

in $\mathcal{S}_{np}$ have the same spectrum $\{\lambda_j\}_{j=1}^{np}$. Equivalently, for any positive integer k and any $s \in \mathcal{S}_{p}$, we have

$$\mathrm{tr}\left(\Sigma_Y^{1/2}(W \otimes s)\Sigma_Y^{1/2} \right)^k = \mathrm{tr}\left(\mathrm{diag}[I_{m_1} \otimes \Sigma_*, -I_{m_2} \otimes \Sigma_*, 0] \right)^k,$$

which proves the equivalence between (c) and (d) via appropriate Kronecker operations. So the proof is complete.

Lemma 7.2: Assume that C-conditions holds. Then there exists an orthogonal matrix H, not depending on i, such that

$$B_{ii} = H(e_{ii} \otimes A_{ii})H', \tag{A.1}$$

where

$$A_{ii} = \mathrm{diag}[U_{ii}, V_{ii}, 0] \in \mathcal{M}_{n \times n}$$

and

$$U_{ii} = I_{m_1}, V_{ii} = -I_{m_2}$$

and

$$B_{ij} = H(e_{ij} \otimes A_{ij} + e_{ji} \otimes A_{ji})H'/2, \tag{A.2}$$

where

$$A_{ij} = diag[U_{ij}, V_{ij}, \mathbf{0}] \in \mathcal{M}_{n \times n},$$

$$U_{ij} \in \mathcal{M}_{m_1 \times m_1},$$

$$V_{ij} \in \mathcal{M}_{m_2 \times m_2}$$

and

$$A'_{ij} = A_{ji}, \; U_{ij}U'_{ij} = I_{m_1},$$

$$V_{ij}V'_{ij} = I_{m_2}.$$

Proof: It follows from condition (C2) that

$$B_{ii}^{2k+1} = B_{ii}, \; \mathrm{tr}\left(B_{ii}^{2k+1}\right) = m_1 - m_2, \; \mathrm{tr}\left(B_{ii}^{2k}\right) = m_1 + m_2, \; k = 1, 2, \ldots \qquad (A.3)$$

By (A.3) and (C3) we may choose an orthogonal matrix H, not depending on i, such that (A.1) holds. Thus using (C3), (C4) and (A.3) it is easily shown that for i≠j, $||B_{ij} - 4B^3_{ij}||^2 = 0$ and so

$$B_{ij} = 4B_{ij}^3, \quad i \neq j. \qquad (A.4)$$

Combining (C4) with (A.4) we obtain, for i≠j, $(B^2_{ii} + B^2_{jj}) B_{ij} = 4B^2_{ij}B_{ij} = B_{ij}$. The symmetry of B_{ij} then yields

$$B_{ij} = (B_{ii}^2 + B_{jj}^2)B_{ij}(B_{ii}^2 + B_{jj}^2), \quad i \neq j. \qquad (A.5)$$

For i≠j, using $e_{ii} \otimes A_{ii}$, $e_{jj} \otimes A_{jj}$ and $H'B_{ij}H$ to replace B_{ii}, B_{jj} and B_{ij}, we get (A.2) from (A.5), which completes the proof.

The following lemma is due to Masaro and Wong [11] and its proof is also found in Hu [6], Appendix.

Lemma 7.3: Let $Y \sim \mathcal{N}_{n \times p}(0, \Sigma_Y)$ with $\Sigma_Y \in \mathcal{N}_{np}$ and W be a symmetric matrix of order n. Then $YWY \sim \mathcal{W}_p(m_1, \Lambda) - \mathcal{W}_p(m_2, \Lambda)$ with $m_1, m_2 \in \{0, 1, ...\}$ and $\Lambda = \mathrm{diag}\,[\Sigma_1, ... , \Sigma_r, 0] \in \mathcal{N}_p$ if and only if there exist positive real numbers $\Sigma_1, \Sigma_2, ... , \Sigma_r (r \pounds\, p)$ such that C-conditions hold.

REFERENCES

1. T.W. Anderson, G.P.H. Styan, Cochran's theorem, rank additivity and tripotent matrices, in: G. Kallianpur (Ed.), et al., Statistics and Probability; Essays in Honor of C.R. Rao, Amsterdam, North-Holland, 1982, pp. 1–23.
2. B. Baldessari, The distribution of a quadratic form of normal random variables, Ann. Math. Statist. 38 (1967) 1700–1704.
3. W.G. Cochran, The distribution of quadratic forms in a normal system with applications to the analysis of covariance, Proc. Cambridge. Philos. Soc. 30 (1934) 178–191.
4. F.J. Graybill, Introduction to Matrices with Applications in Statistics, Wiley, New York, 1969.
5. J. Gurland, Distribution of definite and indefinite quadratic forms, Ann. Math. Statist. 26 (1955) 122–127.
6. J. Hu, Some Extensions of Cochran's Theorem, The Dissertation for the Degree of Doctor of Philosophy at the University of Windsor, Canada, 2007.
7. J. Hu, Wishartness and independence of matrix quadratic forms in a normal random matrix, J. Multivariate Anal. 99 (2008) 555–571.
8. C.G. Khatri, Quadratic forms and extension of Cochran's theorem to normal vector variables, Multivariate Analysis, North-Holland, Krishnaiah, 1977, pp. 79–94.
9. N.Y. Luther, Decomposition of symmetric matrices and distributions of quadratic forms, Ann. Math. Statist. 36 (1965) 683–690.
10. J.R. Magnus, H. Neudecker, The commutation matrix: some properties and application, Ann. Statist. 7 (1979) 381–394.
11. J. Masaro, C.S. Wong, Laplace–Wishart distributions associated with matrix quadratic forms, in: Hawaii International Conference on Statsitics, Mathematics and Related Fields, January 16–19, 2006.
12. J. Masaro, C.S. Wong, Wishartness of quadratic forms: a characterization via Jordan algebra representations, Tamsui Oxf. J. Math. Sci. 25 (2009) 87–117.
13. A.M. Mathai, on noncentral generalized Laplacianness of quadratic forms in normal variables, J. Multivariate Anal. 46 (1993) 239–246.
14. R.J. Muirhead, Aspects of Multivariate Statistical Theory, Wiley, New York, 1982.
15. K. Pearson, S.A. Stuffer, F.N. David, Further application in statistics of the $T_{m(x)}$ Bessel functions, Biometrika 24 (1932) 293–350

16. S.J. Press, Linear combinations of non-central chi-squared variables, Ann. Math. Statist. 37 (1966) 480–487.

17. S.B. Provost, on the exact distribution of the ratio of a linear combination of chi-squared variables over the root of a product of chi-square variables, Canad. J. Statist. 14 (1986) 61–67.

18. C.R. Rao, Linear Statistical Inference and its Applications, Wiley, New York, 1973.

19. J. Robinson, The distribution of a general quadratic form in normal variables, Austral. J. Statist. 7 (1965) 110–114.

20. B.K. Shah, Distribution of definite and indefinite quadratic forms from a noncentral normal distribution, Ann. Math. Statist. 34 (1963) 186–190.

21. W.Y. Tan, Some matrix results and extensions of Cochran's theorems, SIAM J. Appl. Math. 28 (1975) 547–554, Errata 30, 608–610.

22. W.Y. Tan, on the distribution of quadratic forms in normal random variables, Canad. J. Statist. 5 (1977) 241–250.

23. C.S. Wong, J. Masaro, T. Wang, Multivariate versions of Cochran theorems, J. Multivariate Anal. 39 (1991) 154–174.

24. C.S. Wong, T. Wang, Laplace–Wishart distributions and Cochran theorems, Sankhy ̄a Ser. A 57 (1995) 342–359.

CITATION

Jianhua Hu, Equivalent conditions for noncentral generalized Laplacianness and independence of matrix quadratic forms, Linear Algebra and its Applications, Volume 433, Issue 4, 1 October 2010, Pages 796-809, ISSN 0024-3795, http://dx.doi.org/10.1016/j.laa.2010.04.010.

Laplacian Matrices of Graphs: a Survey

Russell Merris
Department of Mathematics and Computer
Science California State University Hayward,
California 94542, USA

ABSTRACT

Let G be a graph on n vertices. Its Laplacian matrix is the n-by-n matrix $L(G)=D(G)-A(G)$, where $A(G)$ is the familiar $(0,1)$ adjacency matrix, and $D(G)$ is the diagonal matrix of vertex degrees. This is primarily an expository article surveying some of the many results known for Laplacian matrices. Its six sections are: Introduction, The Spectrum, The Algebraic Connectivity, Congruence and Equivalence, Chemical Applications, and Immanants.

INTRODUCTION

Let $G = (V, E)$ be a graph with vertex set $V = V(G) = \{v_1, v_2, \ldots, v_n\}$ and edge set $E = E(G) = \{e_1, e_2, \ldots, e_m\}$. For each edge $e_j = \{v_i, v_k\}$, choose one of v_i, v_k to be the positive "end' of ej and the other to be the negative "end." Thus G is given an orientation [ll]. The vertex-edge incidence matrix (or "cross-linking matrix" [33]) afforded by an orientation of G is the n-by-m matrix $Q = Q(G) = (q_{ij} \cdot l$, where $q_{ij} = +1$ if v_1 is the positive end of e_j, - 1 if it is the negative end , and 0 otherwise.

It turns out that the Laplacian matrix, $L(G) = QQt$, is independent of the orientation. In fact, $L(G) = D(G) - A(G)$, where $D(G)$ is the diagonal matrix of vertex degrees and $A(G)$ is the $(0,1)$ adjacency matrix. One may also describe $L(G)$ by means of its quadratic form

$$xL(G)x^t = \sum (x_i - x_j)^2,$$

where $x = (x_1, x_2, ..., x_n)$, and the sum is over the pairs $i < j$ for which $(v_i, v_j) \in E$. So $L(G)$ is a symmetric, positive semidefinite, singular M-matrix.

We are primarily interested in nondirected graphs without loops or multiple edges. However, many of the results we discuss have extensions to edge weighted graphs. A C-edge-weighted graph, G_c, is a pair consisting of a graph G and a positive real-valued function C of its edges. The function C is most conveniently described as an n-by-n, symmetric, nonnegative matrix $C = (c_{ij})$ with the property that $c_{ij} > 0$ if and only if $(v_i, u_j) \in E$. With r_i denoting the ith row sum of C, define $L(G_c) = \mathrm{diag}(r_1, r_2, \ldots, r_n) - C$. Another way to describe $L(G_c)$ is by means of its quadratic form:

$$xL(G_C)x^t = \sum c_{ij}(x_i - x_j)^2,$$

where, as before, the sum is over the pairs $i < j$ for which $\{v_1, v_2\} \in E$.

Forsman [47] and Gutman [66] have shown how the connection between $L(G)$ and $K(G) = Q^tQ$ simultaneously explains the statistical and dynamic properties of flexible branched polymer molecules. Unlike its vertex counterpart, the entries of $K(G)$ depend on the orientation. However, if G is bipartite, an orientation can always be chosen so that $K(G) = 21_m + A(G^*)$, where G^* is the line graph of G. (It follows that the minimum eigenvalue of $A(G^*)$ is at least -2. This observation, first made by Alan Hoffman, leads to a connection with root systems [29, 1451.)

One may view $K(G)$ as an edge version of the Laplacian. For graphs without isolated vertices, there are other versions, e.g., the doubly stochastic matrix $_nI - d_1^{-1}L(G)$, where d_1, is the maximum vertex degree, and the correlation matrix $M(G) = D(G)^{-1/2}L(G)D(G)^{-1/2}$. (A symmetric positive semidefinite matrix is a correlation matrix if each of its diagonal entries is 1.) Note that $M(G)$ is similar to $D(G)^{-1}L(G) = I_n$-$R(G)$, where $R(G)$ is the random-walk matrix. The first recognizable appearance of $L(G)$ occurs in what has come to be known as Kirchhof's matrix tree theorem [77]:

Theorem 1.1: Denote by $L(i|j)$ the $(n - I)$-by-$(n - 1)$ submatrix of $L(G)$ obtained by deleting its ith row and jth column. Then $(-1)^{i+j} \det L(i/j)$ is the number of spanning trees in G. (Variations, extensions and generalizations of Theorem 1.1 appear, e.g., in [8, 16, 17, 25, 26, 51, 78, 94, 97, 127, 132, 149, 150].)

In view of this result, it is not surprising to find $L(G)$ referred to as a Kirchhof matrix or matrix of admittance (admittance = conductivity, the reciprocal of impedance). Reflecting its independent discovery in other contexts, $L(G)$ has also been called an information matrix [25], a Zimm matrix [47], a Rouse-Zimm matrix [130], a connectivity matrix [35], and a vertex-vertex incidence matrix [I53]. Perhaps the best place to begin is with a justification of the name "Laplacian matrix."

In a seminal article, Mark Kac posed the question whether one could "hear the shape of a drum" [74, 115]. Consider an elastic plane membrane whose boundary is fixed. If small vibrations are induced in the membrane, it is not unreasonable to expect a point (x, y, z) on its surface to move only vertically. Thus, we assume $z = z(x, y, t)$. If the effects of damping are ignored, the motion of the point is given (at least approximately) by the wave equation

$$\nabla^2 z = z_{tt} / c^2$$

where $\Delta^2 z = z_{xx} + z_{yy}$ is the Laplacian of .z. Since we are assuming the membrane is elastic and the vibrations are small, the restoring force is linear (Hooke's law), i.e., $z_{tt} = -kz$, where $k > 0$ encompasses mass and "spring constant." Combining these equations, we obtain

$$z_{xx} + z_{yy} = -kz / c^2 \tag{1}$$

The classical solution to this "Dirichlet problem" involves a countable sequence of eigenvalues that manifest themselves in audible tones. An alternate version of Kac's question is this: can nonisometric drums afford the same eigenvalues? (The recently announced answer is yes [146, 148].)

To produce a finite analog, suppress the variable t and use differential approximation to obtain the estimates

$$z(x - h, y) \doteq z(x, y) - z_x(x, y)h,$$
$$z(x, y) \doteq z(x + h, y) - z_x(x + h, y)h.$$

Subtracting the second of these equations from the first and rearranging terms, we find

$$h[z_x(x + h, y) - z_x(x, y)] \doteq z(x + h, y) + z(x - h, y) - 2z(x, y). \tag{2}$$

Another approximation by differentials leads to

$$z_x(x + h, y) \doteq z_x(x, y) + z_{xx}(x, y)h.$$

Putting this into (2) gives

$$h^2 z_{xx} \doteq z(x + h, y) + z(x - h, y) - 2z(x, y)$$

Similarly,

$$h^2 z_{yy} \doteq z(x + h, y) + z(x - h, y) - 2z(x, y)$$

Substituting these estimates into (1), we obtain

$$4z(x, y) - z(x + h, y) - z(x - h, y) - z(x, y + h) - z(x, y - h) \doteq \lambda z(x, y), \tag{3}$$

where $\lambda = kh^2/c^2$. But (3) is the equation $L(G)z = \lambda Z$, where G is the "grid graph" of Figure 1. So the eigenvalue problem for $L(G)$ is, arguably at least, a finite analog of the continuous problem (1). (M. E. Fischer suggested that discrepancies between discrete models like (3) and continuous models like (1) may well reflect the "lumpy nature of physical matter" [46].) The first examples of nonisomorphic graphs G_1 and G_2 such that $L(G_1)$ and $L(G_2)$ have the same spectra were found in [31, 69, 147]. In fact, as we will see in Theorem 5.2, below, there is a plentiful supply of nonisomorphic, Laplacian cospectral graphs.

THE SPECTRUM

Strictly speaking $L(G)$ d e p en d s not only G but on some (arbitrary) ordering of its vertices. However, Laplacian matrices afforded by different vertex orderings of the same graph are permutation-similar. Indeed, graphs G_1 and G_2 are isomorphic if and only if there exists a permutation matrix P such that

$$L(G_2) = P^t L(G_1) P. \tag{4}$$

Thus, one is not so much interested in $L(G)$ as in permutation-similarity invariants of $L(G)$. Of course, two matrices cannot be permutation-similar if they are not similar, and two real symmetric matrices are similar if and only if they have the same eigenvalues.

Denote the spectrum of $L(G)$ by

$$S(G) = (\lambda_1, \lambda_2, ... \lambda_n),$$

where we assume the eigenvalues to be arranged in nonincreasing order: $\lambda_1 \geq \lambda_2 \geq *** \geq \lambda_n = 0$. When more than one graph is under discussion, we may write $\lambda_i(G)$ instead of λ_i. It follows, e.g. from the matrix-tree theorem, that the rank of $L(G)$ is $n - w(G)$, where $w(G)$ is the number of connected components of G. In particular, $\lambda_{n-1} \neq 0$ if

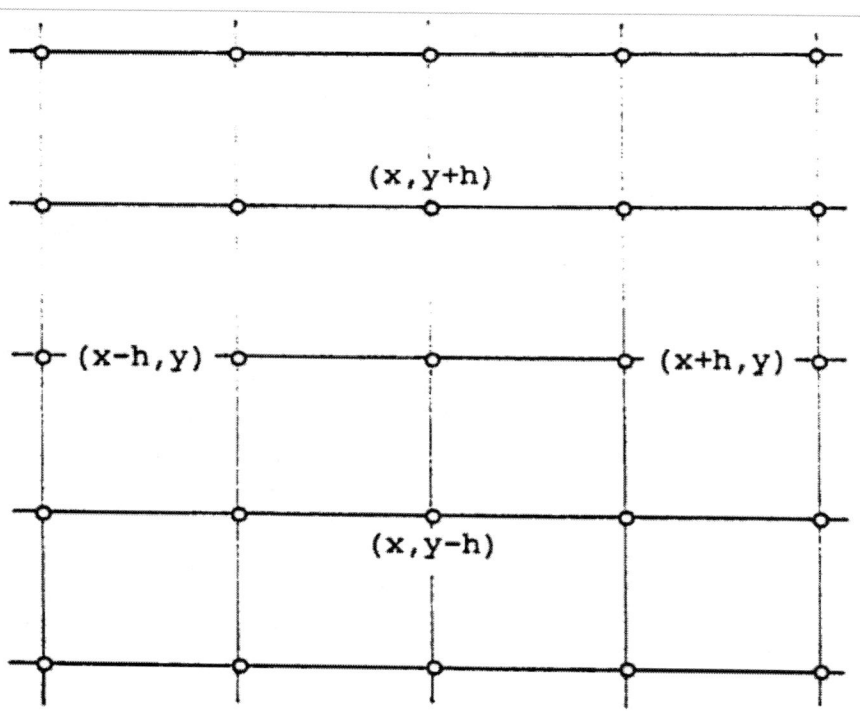

Figure 1: Grid graph.

and only if G is connected. (Already, we see graph structure reflected in the spectrum.) This observation led M. Fiedler [37, 40-43] to define the algebraic connectivity of G by $a(G) = \lambda_{n-1}(G)$, viewing it as a quantitative measure of connectivity. In the next section we will discuss the algebraic connectivity and some of its many applications.

Denote the complement of G (in K_n) by G^c, and let J_n be the n-by-n matrix each of whose entries is 1. Then, as observed in [5], $L(G) + L(G^c) = L(K_n) = nI_n - J_n$, It follows that

$$S(G^c) = (n - \lambda_{n-1}(G), n - \lambda_{n-2}(G),...,n - \lambda_1(G), 0).$$

$$(5)$$

Letting $m_G(\lambda)$ denote the multiplicity of λ as an eigenvalue of $L(G)$, one may deduce from (5) that $\lambda_1(G) \leq n$ and $m_G(n) = w(G^c) - 1$. (See [65] for another interpretation.)

In Section 1, we defined $D(G)$ to be the diagonal matrix of vertex degrees. We now abuse the language by also using $D(G)$ to denote the nonincreasing degree sequence

$$D(G) = (d_1, d_2, ..., d_n),$$

$D_1 \geq d_2 \geq \geq d_n$ (We do not necessarily assume that $d_i = d(v_i)$, the degree of vertex i.) It follows from the GerEgorin circle theorem [applied to $K(G)$] that $\lambda_1 \leq \max[d(u)+d(v)]$, where the maximum is taken over all $\{u, v\} \in E$. (Also see [5].) In particular,

$$d_1 + d_2 \geq \lambda_1.$$

[Note that (6) im . p roves the bound $2d_1 \geq \lambda_1$ obtained by applying Gerzgorin's theorem directly to $L(G)$.]

If $(a) = (a_1\, a_2, ...a_r)$ and $(b) = (b_1\, b_2, ..., b_s)$ are nonincreasing sequences of real numbers, then (a) majorizes (b) if

$$\sum_{i=1}^{k} a_i \geq \sum_{i=1}^{k} b_i, \quad k = 1, 2, ..., \min\{r, s\},$$

and

$$\sum_{i=1}^{k} a_i = \sum_{i=1}^{k} b_i$$

Theorem 2.1: For any graph G, S(G) mujorizes D(G).

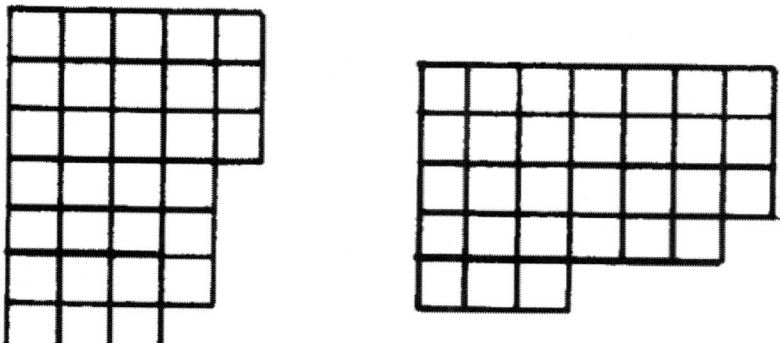

Figure 2: Ferrer-Sylvester diagrams.

Proof

It was proved in [125] (see, e.g., [84, p. 218]) that the spectrum of a positive semidefinite Hermitian matrix majorizes its main diagonal (when both are rearranged in nonincreasing order).

Majorization techniques have been widely used in graph-theoretic investigations ranging from degree sequences to the chemical "Balaban index." (See, e.g., [121, 1221.) I n 1 't s intersection with algebraic graph theory, this work has often been impeded by a stubborn reliance on the adjacency matrix. (See, e.g., [loo].) In fact, it is the Laplacian matrix that affords a natural vehicle for majorization.

The first inequality arising from Theorem 2.1 is $\lambda_1 \geq d_1$. It is not surprising that a result holding for all positive semidefinite Hermitian matrices should be subject to some improvement upon restriction to the class of Laplacian matrices. Indeed [62], if G has at least one edge, then

$$\lambda_1 \geq d_1 + 1. \tag{7}$$

For G a connected graph on n > 1 vertices, equality holds in (7) if and only if $d_1 = n - 1$. In fact, (7) is the beginning of a chain of inequalities that include $\lambda_1 + \lambda_2 \geq d_1 + d_2 + 1$ and $\lambda_i + \lambda_2 + \lambda_3 \geq d_1 + d_2 + d_3 + 1$. These suggest the following:

Conjecture 2.2 [21]: Let G be a connected graph on n ≥ 2 vertices. Then the sequence $(d_1 +1, d_2, d_3,...,d_{n-1}, d_n-1)$ is majorized by S(G).

Nonincreasing integer sequences are frequently pictured by means of so-called Ferrers-Sylvester (or Young) diagrams. For example, the diagram for (a) = (5, 5, 5, 4, 4, 4, 3) is pictured on the left in Figure 2. Its transpose is the diagram on the right corresponding to the conjugate sequence (a)* = (7, 7, 7, 6, 3) In general, the conjugate of a nonincreasing integer sequence (a) = $(a_1, a_2,..., a_r)$ is (a)* = $(a_1^*, a_2^*,..., a_s^*)$, where UT is the cardinality of the set {j:aj ≥ i}.

Theorem 2.3 [62]: Let D(G) be the degree sequence of u graph. Then D(G) mujorizes D(G).

Theorems 2.1 and 2.3 raise the natural question whether S(G) and D(G)* are majorization-comparable.

Conjecture 2.4 [62]: Let G be a connected graph. Then D(G)* majorizes S(G).

One consequence of Conjecture 2.4 would be

$$\lambda_{n-1} \geq d_{n-1}^*, \tag{8}$$

i.e., the number of vertices of G of degree n-1 is no larger than the algebraic connectivity, a(G). Since $a(K_n) = n$, (8) is true for $G = K_n$. Otherwise, if G has exactly k vertices of degree n - 1, then G'' has at least k + 1 components, the largest of which has at most n - k vertices, so $\lambda_1(G^c) \leq n - k$ and $a(G) = n - \lambda_1(G^c) \geq k = d_{n-1}^*$.

There is, of course, an enormous literature on the adjacency spectra of graphs, and much of it concerns regular graphs. (See, e.g., [28-301.) If G is r-regular, $L(G) + A(G) = rI_n$, so λ is an eigenvalue of L(G) if and only if r - λ is an eigenvalue of A(G). Similarly, since L(G) and its edge counterpart, K(G), share the same nonzero eigenvalues, any results about the adjacency spectra of line graphs of bipartite graphs can be carried over to the Laplacian by means of the equation $K(G) = 2I_m + A(G^*)$. These connections with the adjacency literature lead easily to many results for the Laplacian that we won't even try to describe here.

There are some other results about A(G) whose Laplacian counterparts do not follow for the reasons just given, but whose proofs consist of relatively straightforward modifications of adjacency arguments. Three results of this type are presented in Theorems 2.5-2.7.

Theorem 2.5: Let G be a connected graph with diameter d. Suppose L(G) has exactly k distinct eigenvulues. Then $d + 1 \le k$.

Let $\Gamma(G)$ denote the automorphism group of G, regarded as a group of permutations on $V = \{v_1, v_2,..., v_n\}$.

Theorem 2.6: Let G be a connected graph. If some permutation in Γ (G) has s odd cycles and t even cycles, then L(G) has at most $s + 2t$ simple eigenvalues.

If some permutation in T(G) has a cycle of length 3 or more, we see immediately from Theorem 2.6 that the eigenvalues of L(G) are not distinct; if the eigenvalues of L(G) are all distinct, then I'(G) must be Abelian (as each of its elements has order 2).

Denote by $V_1, V_2..., V_t$, the orbits of Γ (G) in V, and let $n_i = o(V_i)$ be the cardinality of V_i, $1 \le i \le t$. Assume V ordered so that

$$V_1 = \{v_1, v_2,..., v_{n_1}\},$$
$$V_2 = \{v_{n_1+1}, v_{n_1+2},..., v_{n_1+n_2}\},$$

etc. Partitioning L(G) in the same way, we obtain a t-by-t block matrix (L_{ij}), where L_{ij} is the n_1-by-n_j submatrix of L(G) whose rows correspond to the vertices in V_i, and whose columns are indexed by the vertices in V_j.

Theorem 2.7 [58]: Let $L(G) = (L_{ij})$ be the block matrix partitioned by T(G) as described above. Let $A = (a_{ij})$ be the t-by-t matrix defined by $a_{ij} = (n_i.n_j)^{-1/2}$ times the sum of the entries in L_{ij}. Then the characteristic pzlynomiai of A is a factor of the characteristic polynomial of L(G).

The eigenvalues of the matrix A in Theorem 2.7, multiplicities included, constitute the symmetric part of the spectrum of L(G). The remaining eigenvalues of L(G), multiplicities included, constitute the alternating part. If $T(G) = \{e\}$, then the aiternating part of the spectrum is empty. On the other hand, it may happen that some multiple eigenvalue of L(G) belongs to both parts.

We now discuss some results directly relating S(G) to various structural properties of G.

Theorem 2.8 [62]: Let u be a cut vertex-of the connected graph G. if the largest component of G-u contains k vertices, then $k + 1 \geq \lambda_2(G)$.

A pendant vertex of G is a vertex of degree 1. A pendant neighbor is a vertex adjacent to a pendant vertex. We suppose G has p(G) pendant vertices and q(G) pendant neighbors.

Theorem 2.9 [36]: Let G be a graph. Then $p(G) - q(G) \leq m_G(1)$.

See Theorem 6.1 (below) for the permanental analog of this result. Extensions of Theorem 2.9 can be found in [59]. The correlation between $m,(l)$ and the viscosity of polydimethylsiloxane is discussed in [llO]. If I is an interval of the real line, denote by $m_G(l)$ the number of eigenvalues of L(G), multiplicities included, that belong to 1. Then $m_G(l)$ is a natural extension of $m_G(\lambda)$.

Theorem 2.10 [63]: Let G be a graph. Then $q(G) \leq m_G[O, 1)$.

It is immediate from Theorems 2.9 and 2.10 that $p(G) \leq m_G[O, 1]$. (The relevance of $m_G(O, 1)$ to long relaxation times in elastic networks is discussed in [11O, p. 885; 130, p. 5184]. Also see [35, Section J].)

Theorem 2.11 [96]: Let G be a connected graph satisfying $2q(G) < n$. Then $q(G) \leq m_G(2, n]$.

A subset S of V(G) is said to be stable or independent if no two vertices of S are adjacent. The maximum size of an independent set is called the interior stability number or the point independence number and is denoted by $\alpha(G)$.

Theorem 2.12: Let G be a graph. Then $m_G[d_n, n] \geq \alpha(G)$ and $m_{G'}[0,1] \geq \alpha(G)$.

Proof. We require the following well-known fact from matrix theory: Suppose that Z3 is a principal submatrix of the symmetric matrix A. Then the number of nonnegative (respectively, nonpositive) eigenvalues of B is a lower bound for the number of nonnegative (respectively, nonpositive) eigenvalues of A. Suppose $S = \{v_1, v_2, ..., v_k\}$ is an independent set of vertices. Let B be the leading k-by-k principal submatrix of $L(G)-d_{n'}I_{n'}$. Then B is a diagonal matrix, each of whose eigenvalues is nonnegative. Therefore, k is a lower bound for the number of nonnegative eigenvalues of $L(G) - d_{n'} I_{n'}$. The argument for $m_G[O, d_1]$ is similar.

If G is r-regular, then Theorem 2.12 becomes

$$m_G[r,n] \geq \alpha(G) \leq m_G[0,r],$$

from which one may recover the regular case of an analogous result for the adjacency matrix [30, Theorem 3.141.

Theorem 2.13 [63]: if T is a tree with diameter d, then $m_T(O,2) \geq [d/2]$, the greatest integer in d/2, and $m_T(2, n] \geq [d/2]$.

It follows, of course, that $m_{P_n}(2) = 1$ if and only if n is even. In fact [63, Theorem 2.5], $m_T(2) = 1$ for any tree T with a perfect matching.

Theorem 2.14 [62]: Let G be a graph. If $m_G(2) > 0$, then $d(u) + d(v) \leq n$ for some pair of nonadjacent vertices u and v.

Theorem 2.15: Let G be a connected graph. If t is the length of a longest path in G, then $m_G(2, n] \geq [t/2]$

Proof: If G is a tree, then t is the diameter and we use Theorem 2.13. Otherwise, the longest path in G is part of a spanning tree T. Since G may be obtained from T by adding edges, the result follows from Theorem 2.16.

The next result is part of the "Laplacian folklore" [63, 1041.

Theorem 2.16: If u and w are nonadjacent vertices of G, let G^+ be the graph obtained from G by adding a new edge e = {u, w}, Then the n-1 largest eigenvalues of L(G) interlace the eigenvalues of $L(G^+)$.

If u $\in$ V, denote by N(u) its set of neighbors, i.e.,

$N(u) = \{v \in V : \{u, v\} \in E\}$.

[If X $\subset$ V, then N(X) is the union over u $\in$ X of N(u).]

Wasin So [131] found a nice addition to Theorem 2.16: If N(u) = N(w), then the spectrum of $L(G^+)$ overlaps the spectrum of L(G) in n - 1 places. That is, in passing from L(G) to $L(G^+)$, one of the eigenvalues goes up by 2 and the rest are unchanged.

Theorem 2.17 [63]: If T is a tree and λ is any eigenvalue of L (T), then $m_l(h) \leq p(T) - 1$.

Recall that p(T) - 1 is also an upper bound for the nullity of A(T) [30, p. 2581. If G IS connected and bipartite, then L(G) = D(G) - A(G) is unitarily similar to the irreducible nonnegative matrix D(G) + A(G), and λ_1 (G) is a simple eigenvalue.

Theorem 2.18 [63, Theorem 2.11: Suppose T is tree. If $\lambda > 1$ is an integer eigenvalue of L(T) with corresponding eigenvector u, then λ I n, $m_T(\lambda) = 1$, and no coordinate of u is 0.

This may be a good time to recall a striking result of Fiedler [38]: Suppose A = 2 is an eigenvalue of L(T) for some tree T = (V, E). Let z = $(z_1, z_2, \ldots, z_n)$ be an eigenvector of L(T) corresponding to λ = 2. Then the number of eigenvalues of L(T) greater than 2 is equal to the number of edges $\{v_i, v_j\} \in E$ such that $z_i z_j > 0$.

Let $G_1 = (V_1, E_1)$ and $G_2 = (V_2, E_2)$ be graphs on disjoint sets of vertices. Their union is $G_1 + G_2 = (V_1 \cup V_2, E_1 \cup E_2)$. A coalescence of G_1 and G_2 is any graph on $o(V_1) + o(V_2) - 1$ vertices obtained from $G_1 + G_2$ by identifying (i.e., "coalescing" into a single vertex) a vertex of G_1

with a vertex of G_2. Denote by $G_1 \bullet G_2$ any of the $o(V_1)o(V_2)$ coalescences of G_1 and G_2.

Theorem 2.19 [61]: Let G_1 and G_2 be graphs. Then $S(G, . G,)$ major&s $S(G_1 + G_2)$.

The join, $G_1 \vee G_2$ of G_1 and G_2 is the graph obtained from $G, + G,$ by adding new edges from each vertex of G_1 to every vertex of G_2 Thus, for example, $K_p^c \vee K_q^c = K_{p,q}$ the complete bipartite graph. Because $G_1 \vee G_2 = (G_1^c + G_2^c)^c$, the next result is an immediate consequence of (5):

Theorem 2.20: Let G_1 and G_2 be graphs on n_1 and n_2 vertices, respectively. Then the eigenvalues of $L(G_1 \vee G_2)$ are 0; $n_1 + n_2$; $n_2 + \lambda_i(G_1)$ 1 $\le i < n_1$; and $n_1 + \lambda_i(G_2)$, $1 \le i < n_2$.

The product of G_1 and G_2 is the graph $G_1 \times G_2$ whose vertex set is the Cartesian product $V(G_1) \times V(G_2)$. Suppose $v_1, v_2 \in V(G_1)$ and $u_1, u_2 \in V(G_2)$. Then (v_i, u_i) and $(v_2 u_2)$ are adjacent in $G_1 \times G_2$ if and only if one of the following conditions is satisfied: (i) $v_1 = v_2$ and $\{u_1, u_2\} \in E(G_2)$, or (ii) $\{v_1, v_2\} \in E(G_1)$ and $u_1 = u_2$. For example, the line graph of $K_{p,q}$ is $K_p \times K_q$, and the "grid graph" is a product of paths.

Theorem 2.21 [37, 104]: Let G_1 and G_2 be graphs on n_1 and n_2 vertices, respectively. Then the eigenvalues of $L(G_1 \times G_2)$ are all possible sums $\lambda_i(G_1) + \lambda_j(G_2)$, $1 \le i \le n_1$, and $1 \le j \le n_2$.

Majorization results involving products can be found in [24].

The study of graphs whose adjacency spectra consist entirely of integers was begun in [68]. Cvetkovic [27] p roved that the set of connected, r-regular adjacency integral graphs is finite. When r = 2 there are three such graphs, C_3 C_4 and C_6 when r = 3 there are 13 [15, 129]. Of course, the theory of Laplacian integral graphs coincides with its adjacency counterpart for regular graphs. Elsewhere, there can be remarkable differences. Of the 112 connected graphs on n = 6 vertices, six are adjacency integral while 37 are Laplacian integral.

It is clear from (5) that the spectrum of L(G) consists entirely of integers if and only if the spectrum of $L(G^c)$ is integral. From Theorems 2.20 and

2.21, we see that joins and products of Laplacian integral graphs are Laplacian integral. If T is a tree, then $a(T) < 1$ unless $T = K_{1,n-1}$ [37, 93]. Since $s(K_{1,n-1}) = (n, 1, 1,..., 1, O)$, the star is the only Laplacian integral tree on n vertices. Additional results on Laplacian integral graphs can be found in [62]. We conclude this section with a pair of results that guarantee the existence of certain particular integers in S(G).

A cluster of G is an independent set of two or more vertices of G, each of which has the same set of neighbors. The degree of a cluster is the cardinality of its shared set of neighbors, i.e., the common degree of each vertex in the cluster. An s-cluster is a cluster of degree s. The number of vertices in a cluster is its or&r. A collection of two or more clusters is independent if the clusters are pairwise disjoint. (The neighbor sets of independent clusters need not be disjoint).The next result is an extension of [36].

Theorem 2.22 [154]: Let G be a graph with k independent s-clusters of orders $r_1, r_2, , r_k$. Then $m_G(s) \geq r_1 + r_2 +... + r_k - k$.

Corollary 2.23 [62]: Let G be a graph with an r-clique, $r \geq 2$. Suppose every vertex of the clique has the same set of neighbors outside the clique. Let the degree of each vertex of the clique be s, so $s - r + 1$ is the number of vertices not belonging to the clique but adjacent to every member of the clique. Then $m_G(s + 1) \geq r - 1$.

Proof: The clique corresponds to an $(n - s - 1)$ -cluster of G^c of order r.

THE ALGEBRAIC CONNECTIVITY

Recall that the algebraic connectivity is $a(G) = \lambda_{n-1}(G)$. We begin this section with an early result of Fiedler.

Theorem 3.1 [37]: Let G be a graph (on n vertices) with vertex connectivity $v(G)$ and edge connectivity $e(G)$. Then $2e(G)[1 - \cos(\pi-/n)] \leq a(G)$. If $G \neq K_n$, then $a(G) \leq v(G)$.

If $G \neq K_n$ one deduces that

$$a(G) \le d_n, \tag{9}$$

the minimum vertex degree. An improvement on (9) can be found in [ill]. It seems that $a(G)$ is related to the half-life of a certain "flowing process" in graphs [82]; its relevance to the theory of elasticity is discussed in [13O]. The asymptotic behavior of $a(G)$ f or random graphs is described, e.g., in [73, 87, 111, 1301]. An inequality for the continuous analog of $a(G)$ in compact Riemannian manifolds was obtained by J. Cheeger [18].

Suppose X is a subset of $V(G)$ of cardinality $o(X)$. Define the coboundary, E_X to be the edge cut consisting of those edges exactly one of whose vertices belong to X:

$$E_x = \{\{u, v\} \in E(G) : u \in X \text{ and } v \notin X\}.$$

The isoperimetric number of G is $i(G) = \min[o(E_x)/o(X)]$, where the minimum is over all $X \subset X(G)$ satisfying $1 \le o(X) \le n/2$.

Theorem 3.2 [102, 103]: If G is a graph on $n > 3$ vertices, then

$$\frac{a(G)}{2} \le i(G) \le \{a(G)[2d_1 - a(G)]\}^{1/2}.$$

Related isoperimetric inequalities were established in [5O, 1511, and a continuous analog appeared in [57]. Graphs with large $a(G)$ are related to so-called expanders [2]. (See [3, 4, 10, 19, 20, 50, 104, 108].)

We now state, in terms of Laplacians, a result of M. Doob [30, p. 187].

Theorem 3.3: Let T be a tree on n vertices with diameter d. Then $a(T) \le 2\{1 - \cos[\pi/(d + 1)]\}$

The next result, attributed to B. McKay [108], was proved in [105].

Theorem 3.4: Let G be a connected graph with diameter d. Then a(G) $\geq$ 4/dn.

Another bound involving a(G) and the diameter of G was obtained by Alon and Milman:

Theorem 3.5 [4]: Let G be a connected graph with maximum vertex degree d_1 Then $[2d_1/a(G)]^{1/2} \log_2(n^2)$ is an upper bound for the diameter of G.

Improvements on this result have been obtained by Mohar [105] and Chung, Faber, and Manteuffel [20]. (See [108].)

An upper bound for the diameter in terms of the number of I's in the Smith normal form of L(G) is given in Theorem 4.5 below.

We now consider eigenvectors corresponding to a(G). (These eigen-vectors play an interesting role in the study of random elastic networks 1331 and in the solution of large, sparse, positive definite systems on parallel computers [114].) Denote by Val(G) the set of eigenvectors of L(G) afforded by a(G). Then Val(G) lacks only the zero vector to be a vector space. For our present purposes, it is useful to think of the ele-ments of Val(G) as real-valued functions of V = V(G). If, for example, z = $(z_1, z_2, \ldots, z_n)$ is an eigenvector of L(G) afforded by a(G), we write $f \in$ Val(G) for the function defined by $f(v_i) = z_i$, $1 \leq i \leq n$. Fiedler has called the elements of Val(G) characteristic valuations of G.

Theorem 3.6 [39]: Let T = (V, E) be a tree. Suppose $f \in$ Val CT). Then two cases can occur.

Case (i): If $f(u) \neq$ f 0 f or all $v \in$ V, then T contains exactly one edge (u, w) such that f(u) > 0 and f(w) < 0. Moreover, the values off along any path starting at u and not containing w increase, while the values off along any path starting at w and not containing u decrease. *Case (ii):* If $V_0 = \{u \in V : f(v) = 0\}$ is not empty, then the graph $T_0 = (V_0, E_0)$ induced by T on V_0 is connected and there is exactly one vertex $u \in V$, which is adjacent (in T) to a vertex not belonging to V_0. Moreover, the values off along any path in T starting at u are increasing, decreasing, or identi-cally zero.

Suppose $f \in \text{Val}(T)$. A vertex $v \in V$ is a characteristic vertex of T defined by f if $v \in \{u, w\}$ in case (i), or if $v = u$ in case (ii), whichever applies to f. It turns out that characteristic vertices are independent of the characteristic valuation used to define them: If $f, g \in \text{Val}(T)$, then $v \in V$ is a characteristic vertex of T defined by g if and only if it is a characteristic vertex of T defined by f [93]. Th us, every tree has a unique characteristic center consisting of either one or two characteristic vertices, and in the case of two, they are adjacent. (In spite of these similarities, the characteristic center of a tree need coincide with neither the center nor the centroid.) We say T is of type Z if it has a single characteristic vertex [which must be a fKed point of $I'(T)$]. Otherwise it is of type II. (The algebraic connectivity of a type-I tree is a unit in the ring of algebraic integers [58]. The algebraic connectivity of a type-II tree is a simple eigenvalue of $L(T)$ [38].)

Let T be a type-1 tree with characteristic vertex u_T. A branch at u_T is a connected component of $T\text{-}u_T$. If B is a branch at u_T, denote by $r(B)$ the vertex of B adjacent (in T) to u_T. If $f \in \text{Val}(T)$, then (Theorem 3.6) f is uniformly positive, uniformly negative, or identically zero on the vertices of B. We call B a passive branch if $f(r(B)) = 0$ for every $f \in \text{Val}(T)$. Otherwise, B is active. In either case, denote by $L^+(B)$ the matrix obtained from $L(B)$ by adding 1 to its main-diagonal entry in the row corresponding to $r(B)$. Then the $(n - 1)$-by-$(n - 1)$ principal submatrix of $L(T)$ obtained by deleting the row and column corresponding to ur is the direct sum of the $L^+(B)$ as B ranges over the branches of T at u_T. This leads to the following:

Theorem 3.7 [58]: Let T be a type-I tree with characteristic vertex u_T and algebraic connectivity $a(T)$. Then, for every branch B of uT at I+, $a(T) \leq$ the least eigenvalue of $L^+(B)$, with equality if and only if B is active, in which case $a(T)$ is a simple eigenvalue of $L^+(B)$.

It is a consequence of Theorem 3.7 that exactly $m_T(a(T)) + 1$ of the branches at u_T are active. If $a(T)$ is a simple eigenvalue of $L(T)$, then u_T and the passive branches "separate" the two active branches in the following sense: A subset $C \subset V(G)$ is said to separate vertex sets X and Y if(i) X, Y, and C partition $V(G)$, and ("> u no vertex of X is adjacent to a vertex of Y. It is of some interest to find separators with o(C) small and o(X) about equal to o(Y).

Theorem 3.8 [4, Lemma 2.11: Suppose C separates X and Y in the connected graph G. Let x = o(X), y = o(Y), z be the number of edges having at least one "end" in C, and d be the minimum distance between a vertex in X and a vertex in Y. Then $(x^{-1}+y^{-1}).z/d^2 \geq a(G)$.

See [114] for improvements on this result.

Corollary 3.9: Let T be a type-I tree with characteristic vertex u_T and simple algebraic connectivity a(T). If x and y are the numbers of vertices in the two active branches of T at u_T, then a(T) $\leq$ (x + y)/(2xy).

Proof: Let T' be the subtree induced by T on ur and the two active branches. It is proved in [93] that a(T') = a(T). The result is an immediate consequence of Theorem 3.8.

It is known [58] that a(T) is in the alternating part of the spectrum if and only if at least two of the active branches at ur are isomorphic. If T has just two isomorphic branches at u_T, then a(T) $\leq$ 2/(n - 1) (x = y in Corollary 3.9).

The algebraic connectivity for trees on n vertices ranges from $a(P_n) =$ 2[1 - cos(π/n)] to a($K_{1,n-1}$) = 1. Clearly, then, all trees are not equally connected. Some results explaining the partial ordering imposed on trees by a(T) were obtained in [60]. (Also see [113].) Other approaches appear in Theorems 3.10 and 3.13. The first of these shows that graphs with large a(G) do not contain small separators.

Theorem 3.10 [108]: Supp ose C separates X and Y in the connected graph G. Let x = o(X), y = o(Y), and c = o(C). Then c $\geq$ 4xya(G)/[nd$_1$ - a(G)(x + ~1].

In [104], Mohar investigated a bandwidth-type problem. For each edge e = {v_i,v_j}, he defined jump(e) = |i -j|, and suggested that ordering the vertices, v_1, v_2,..., v_n, by the values of a characteristic valuation comes close to minimizing

$$\text{Jump}(G) = \sum_{e \in E(G)} [\text{jump}(e)]^2.$$

Theorem 3.11 [104]: $\text{Jump}(G) \geq a(G)n(n^2 - 1)/12$.

If T_C is a C-edge weighted tree (see Section I), denote by $a(T_c)$ the second smallest eigenvalue of $L(T_c)$. The absolute algebraic connectivity of the (unweighted) tree T is $\hat{a}(T) = \max a(T_c)$, where the maximum is over all positive real-valued functions C of E(T) that satisfy $C_{i<j}.c_{ij} = n - 1$.

Associated with T is a metric space T_m, obtained by 1 d entifying each edge of T with the unit interval [O, 11. The points of T_m, are the vertices of T together with all points of the (unit interval) edges. The graph-theoretic distance between vertices in T extends naturally to a metric $d(x, y)$ between points x and y in T_m. The variance of T is

$$\text{var}(T) = \min_{x \in T_m} \sum_{v \in V} \frac{d(x,v)^2}{n-1}.$$

Theorem 3.12 [43]: Let T be a tree. Then $\hat{a}(T) = l/\text{var}(T)$.

Let G_C be a C-edge-weighted graph. For $X \subset V(G)$, let E_X, be its co-boundary. Define the max cut of G_C, by

$$MC(G_c) = \max_{X \subset V(G)} \sum_{\{v_i, v_j\} \in E_X} c_{ij}.$$

Theorem 3.13 [107]: Let $\lambda_1(G_c)$ be the maximum eigenvalue of the C-edge weighted Laplacian $L(G_c)$. Then

$$MC(G_c) = \frac{n\lambda_1(G_c)}{4}.$$

The C-edge-weighted Laplacian is distantly related to the positive semidefinite symmetric matrices $B = (b_{ij})$ satisfying $\sum b_{ij} = 1$ and $b_{ij} = 0$ for $\{v_i, v_j\} \in E(G)$ that are used in [Sl] to study the Shannon capacity. Results involving chromatic numbers and multiplicities of eigenvalues of other matrices distantly related to C-edge-weighted Laplacians can be found in [136, 137].

CONGRUENCE AND EQUIVALENCE

As we have seen [Equation (4)], G_1 and G_2 are isomorphic if and only if there is a permutation matrix P such that $P^t L(G_1)P = L(G_2)$. Thus, one necessary condition for two graphs to be isomorphic is that they have similar Laplacian matrices, partially explaining all the interest in the Laplacian spectrum. But there are other ways to view (4). Recall that an n-by-n integer matrix U is unimodulur if $\det U = \pm 1$. So the unimodular matrices are precisely those integer matrices with integer inverses. Two integer matrices A and B are said to be congruent if there is a unimodular matrix U such that $U^t AU = B$. Because permutation matrices are unimodular, another interpretation of (4) is that two graphs are isomorphic only if they have unimodularly congruent Laplacian matrices. Henceforth, we will say G_1 and G_2 are congruent if there is a unimodular matrix U such that $U^t L(G_1)U = L(G_2)$. The first significant work on congruent graphs was done by William Watkins.

Theorem 4.1 [140]: Suppose G_1 and G_2 are graphs on n vertices. If the blocks of G_1 are isomorphic to the blocks of G_2 then G_1 and G_2 are congruent.

Watkins showed that the converse of Theorem 4.1 fails by exhibiting a pair of congruent, nonisomorphic Z-connected graphs. Two graphs, G_1 and G_2 are cycle-isomorphic (or Sisomorphic [144]) if there is a bijection $f : E(G_1) \rightarrow E(G_2)$ with the property that Y is the set of edges constituting a cycle in G_1, if and only if f(Y) is the set of edges constituting a cycle in G_2.

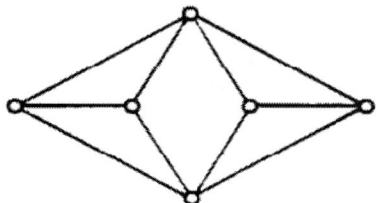 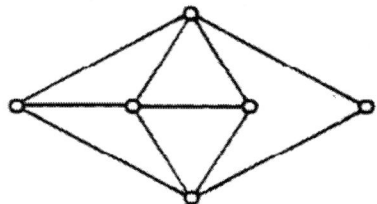

Figure 3: Read's chromatically equivalent graphs.

Theorem 4.2 [141]: Let G_1, and G_2, be graphs with n vertices. Then G_1 and G_2 are congruent if and only if they are cycle-isomorphic.

Denote the chromatic polynomial of G by

$$p_G(x) = \sum_{t-=0}^{n-1} (-1)^t c_t(G) x^{n-t}.$$

(10)

Then $P_G(k)$ is the number of ways to color the vertices of G, using k colors, in which adjacent vertices are colored differently. Using either matroid theory or Whitney's theorem [143], one may easily deduce the following from Theorem 4.2:

Corollary 4.3 [97]: Congruent graphs afford the same chromatic polynomial.

The converse of Corollary 4.3 is false. R. C. Reed [116] produced the pair of "chromatically equivalent" graphs illustrated in Figure 3. Since they have 128 and 120 spanning trees, respectively, they are not even equivalent (see below), much less congruent.

Using another result of Whitney [144], one may draw a potentially more important conclusion from Theorem 4.2:

Corollary 4.4 [141]: If G_1 is a S-connected graph, then G_1 and G_2 are isomorphic if and only if they are congruent.

The fact that there is no canonical form for congruence [49] places some practical limitations on the usefulness of Corollary 4.4. On the other hand, integer matrices cannot be congruent if they are not equivalent, and the question of unimodular equivalence is easily settled by means of the Smith normal form.

Recall that integer matrices A and B are equivalent if there exist unimodular matrices U_1 and U_2 such that $U_1 AU_2 = B$. So a third interpretation of (4) is that G_1 and G_2 are isomorphic only if their Laplacian matrices are equivalent. For the purpose of this article, we will say two graphs are equivalent if their Laplacian matrices are equivalent.

Denote by $d_k(G)$ (not to be confused with vertex degrees) the kth determinantal divisor of L(G), i.e., the greatest common divisor of all the k-by-k determinantal minors of L(G). [It follows from the matrix-tree theorem that $d_{n-1}(G)$ is the number of spanning trees in G; and $d_n(G) = 0$, because L(G) is singular.] Of course, $d_k(G) \mid d_{k+1}(G)$, $0 < k < n$. The invariant factors of G are defined by $s_{k+1}(G) = d_{K+1}(G)/d_K(G)$, $0 \le k < n$, where $d_0(G) = 1$. The Smith normalform of L(G) is

$$F(G) = \text{diag}(s_1(G), S_2(G), ..., s_n(G)).$$

So G_1 and G_2 are equivalent if and only if $F(G_1) = F(G_2)$. In particular, if G_1 and G_2 are isomorphic, then FCC,) = F(G,). Now, if this observation had a partial converse, e.g., for 3-connected graphs, it would have great computational significance because F(G) can be obtained from L(G) by a sequence of elementary row and column operations. However, the graphs in Figure 4 share the Smith normal form diag(1, 1, 1,5,15, O), and the graph on the left ($P_2 \times C_3$) is 3-connected.

In spite of this discouraging example, F(G) yields several bona fide graph-theoretic invariants and spawns a variety of applications: The cycle space, C_G (not to be confused with G_c), of the oriented graph G is the column null space of the vertex-edge incidence matrix Q(G) [and hence the null space of the "edge version" K(G)]; the cocycle space or bond space, R,, is the row space of Q(G).

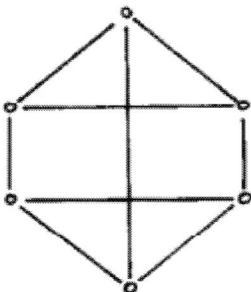 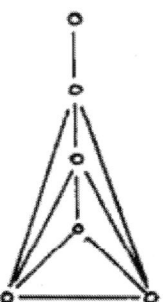

Figure 4: Graphs with the same Smith normal form.

As a subspace of real (or complex) m-space, the "bicycle" space, $B_G = C_G \cap R_G$ is trivially equal to (0). When the coefficients come from Abelian groups, however, one obtains an analogous bicycle group which may be more interesting. K. A. Berman ([9], but see [85] and/or [97] for a clarifying discussion) used the invariant factors of L(G) to completely characterize bicycle groups. Meanwhile, from another perspective, D. J. Lorenzini [79] investigated a similar application of F(G) to the components of the N&on model of the Jacobian associated with a generic curve in algebraic geometry.

Denote by b(G) the multiplicity of 1 in F(G). Of course, $b(G) \geq n - 2$, for any graph G with a square-free number of spanning trees. Lorenzini [80] discusses a bound for b(G) in terms of the number of independent cycles of G.

Theorem 4.5 [64]: Let G be a connected graph of diameter d. Then $b(G) \geq d$.

At the present time, a clear understanding of the relation of the invariant factors, $s_i(G)$, $1 < i < n - 1$, to graph structure seems rather distant. In rather stark contrast, however, the Smith normal form of K(G) has been described completely.

Theorem 4.6 [97]: Let G be a connected graph with n vertices and m > 0 edges. Then the Smith normalform of K(G) is $I_{n-2} + i(n) + O_{m-n+1}$, where the identity (direct) summand is absent when m = 1, and the zero summand is missing when m = n - 1.

Theorem 4.6 has applications to certain "flows" in directed graphs. Of these, the "O-flows," or "A-flows," have been counted by D. Welsh using the chromatic polynomial of the cocycle matroid 11421.

The elementary divisors of L(G) are the prime power factors of its invariant factors. Denote by el(G) the multiset of these elementary divisors.

Theorem 4.7 [97, 140]: Let G_1 G_2 be any coalescence of G_1 and G_2 Then $el(G_1 - G_2) = el(G_1 + G_2) = el(G_1) \cup el(G_2)$.

CHEMICAL APPLICATIONS

Modern organic chemists have synthesized and/or isolated several million different molecules [118]. Perhaps even more remarkable has been their ability to predict certain properties of chemical substances even before they have been synthesized. Among the tools used in such predictions are numerous "topological indices" (a term coined by Haruo Hosoya in 1971 [120]>. A typical t o p o 1 o gi ca 1 in d ex is a number arising from the underlying graph of a chemical compound. (See, e.g., [6, 94, 101, 108, 109, 117-1211.)

The Wiener index, introduced by Harry Wiener of Brooklyn College in 1947, has been used in a variety of ways from predicting antibacterial activity in drugs to correlating thermodynamic parameters in physical chemistry and modeling various solid-state phenomena [67]. It can be obtained by summing the entries in the upper triangular part of the "distance matrix" [l33, p. 451.

The distance, d(u, v), between vertices u and o in a connected graph G is the number of edges in a shortest path from u to v. The distance matrix Δ (G) = $(d(v_i, v_j))$ is the n-by-n matrix whose (i, j) entry is the distance from v_i to v_j. So Δ (G) is a symmetric matrix with zeros along the main diagonal. Hosoya [70, 71] was among the first to study the distance matrix from a chemical perspective. It has since become a standard tool used in a variety of applications from investigating evolutionary distances in DNA sequences to predicting carcinogenicity in arene systems [llQ]. In the mathematical literature, distance matrices seem first to have appeared in [56], where the following remarkable result was proved:

Theorem 5.1: Let T be a tree on n vertices. Then det Δ (T) = $(-l)^{n-1}$ (n - 1) 2^{n-1}

One surprising thing about Theorem 5.1 is that det Δ (T) depends only on n and not at all on the structure of T. In any event, it follows that Δ (T) is an invertible matrix with exactly one positive eigenvalue. In spite of this elegant beginning, results about the distance matrix have not come easily. (See [2l, 22, 32, 54, 55, 123].)

If there were a "Holy Grail" in graph theory, it would be a practical test for graph isomorphism. In the early days, it was incautiously conjectured that two graphs are isomorphic if and only if they have similar adjacency matrices, i.e., that two adjacency matrices could not be similar without being permutation-similar. The disproof of this conjecture began the study of adjacency cospectral graphs: G_1 and G_2 are adjacency cospectral if $A(G_1)$ and $A(G_2)$ have the same characteristic polynomial. One of the most dramatic results in algebraic graph theory is Allen Schwenk's "almost all trees are cospectral" theorem [128]: Let t_n be the number of nonisomorphic trees on n vertices. Let r_n be the number of such trees T for which there exists a nonisomorphic tree T' such that T and T' are adjacency cospectral. Then lim $_{n\to\infty} r_n / t_n = 1$.

Perhaps A(G) is just the wrong matrix. Maybe it is too sparse. What about the distance matrix, whose only zeros occur on the main diagonal? Not surprisingly, it was conjectured that two trees could not be distance-cospectral without being isomorphic [32, 71]. Th' IS conjecture eventually led to the following worthy successor of Schwenk's theorem.

Theorem 5.2 [86]: Let t_n be the number of nonisomorphic trees on n vertices. Let r_n be the number of such trees T for which there exists a nonisomorphic tree T' such that, simultaneously,

(i)   T and T' are adjacency-cospectral,
(ii) T and T' are distance-cospectral, and
(iii) T and T' are Laplacian-cospectral.
Then $\lim_{n\to\infty} r_n / t_n = 1$.

Additional simultaneous conditions will be added in Theorem 6.6. The next result [95] (also see [55]) furth er strengthens the spectral relationship between L(T) and Δ (T).

Theorem 5.3: Let T be a tree. Then the eigenvalues of -2 K(T)$^{-1}$ interlace the eigenvalues of Δ (T). That is, let $\delta_1 > 0 > \delta_{2'} \geq \cdots \geq \delta_{N'}$, be the eigenvalues of Δ (T), and suppose $\lambda_1 \geq \lambda_2 \geq \cdots \geq \lambda_{n-1}$ are the nonzero eigenvalues of L(T). Then

$$0 > \frac{-2}{\lambda_1} \geq \delta_2 \geq \frac{-2}{\lambda_2} \geq \delta_3 \geq \cdots \geq \frac{-2}{\lambda_{n-1}} \geq \delta_n.$$

Theorem 5.3 makes it possible to transcribe some Laplacian spectral results for distance matrices. Suppose, for example, T is a tree with diameter d. Then $\delta_{[d/2]} > -1$. If T has p pendant vertices and q pendant neighbors, then $\delta_q > -1$ n > 2q $\delta_{n-q+2} < -2, \delta_p \geq -2$, and δ_{n-p+2} 2. If δ is an eigenvalue of Δ (T) of multiplicity k, then k ≤ p; among the eigenvalues of Δ (T), δ = -2 occurs with multiplicity at least p - q - 1. (Karen Collins has improved this to p - q [23].)

In a remarkable tour de force, M. Fiedler 1441 placed Theorem 5.3 in a geometrical setting. One may view L(G) as a Gram matrix based on n vectors, $y_1, y_2, \ldots, y_n$ in (n − 1)-dimensional real Euclidean space E_{n-1}. Let S be the unit sphere centered at the origin in E_{n-1}. Let P_i be the hyperplane tangent to S at its intersection with the ray generated by y_i. Let H_1 be the half space, determined by P_1 that contains the origin. Then the intersection of the H produces a simplex. Let $A_1, A_2, \ldots, A_n$, be the vertices of this simplex. Define e_{ij} to be the square of the Euclidean distance (in E_{n-1}) between A_i and A_j. The matrix E(G) = (e_{ij}) is called the (Cayley-)Menger matrix of the simplex [152].

Theorem 5.4 [44, 150]: If T is a tree, then $E(T) = \Delta(T)$, i.e., the distance matrix of a tree is the Merger matrix of the simplex arising from its Laplacian matrix.

Returning to the Wiener index

$$W(G) = \sum_{i<j} d(v_i, v_j),$$

we have the following result [88, 94, 95, 101, 105]:

Theorem 5.5: Let T be a tree with Laplacian eigenvalues $\lambda_1 \geq \lambda_2 \geq \cdots \geq \lambda_{n-1} > 0 = \lambda_n$ Then W(T) is given by

$$\sum_{i=1}^{n-1} \frac{n}{\lambda_i}. \tag{11}$$

For a general graph, the Wiener index is not a function of its Laplacian spectrum [94]. This suggests that one might define $W_1(G) = W(G)$ and $W_2(G)$ by (11). Then $W_1(G) = W_2(G)$ if G is a tree, but the two indices may differ otherwise. Indeed, since the dominant contribution to $W_2(G)$ is n/λ_{n-1} it may even be of interest to study $W_3(G) = n/a(G)$. Other possibilities are suggested in [101]. (Some ideas for computing W(G) are contained in [106].)

The expression (11) has turned up in some other contexts. It is, for example, n2 times the mean squared radius of gyration of a polymer molecule [33-35, 110]. If S(n, m) denotes the set of all graphs on n vertices having at most m edges, then minimizing (11) over S(n, m) corresponds to the A-optimality criterion in statistical design [25, p. 156].

IMMANANTS

Let $A = (a_{ij})$ be an n-by-n matrix. Then

$$per\, A = \sum_{p \in S_n} \prod_{t=1}^{n} a_{tp(t)},$$

(12)

where S_n is the symmetric permutation group. Permanents of adjacency matrices have received some attention, especially by chemists who use them to count "Kekule structures" in chemical graphs [30, 124]. (August Kekule von Stradonitz first proposed the hexagonal configuration for the benzene molecule [l, 133]. Arising from the alternating single and double bonding between carbon atoms in $C_6 H_6$, a Kekule structure corresponds to a perfect matching in the underlying carbon skeleton. The "Char postulate" [134] asserts that a benzenoid system with no Kekule structure should be an unstable biradical. At the same time, aromatic compounds-characterized by benzene-like ring structures-"survive intact over geologic time and even persist in the harsh environment of nebulae" [l].)

The serious study of per $L(G)$ seems to have begun with the conjecture [98] that per $L(G) \geq 2(n-1)$, for all connected graphs on n vertices. This conjecture was established by R. Brualdi and J. L. Goldwasser [14] in the course of their study of the Luplacian ratio, per Πd_i. (Also see [7, 13, 52, 90, 138, 139]. It was suggested in [112] that minper $A = $ per $L(K_n)$, where the minimum is over the set of all singular correlation matrices.)

Now, per $L(G)$ is just the constant coefficient in the Laplacian permunental polynomial

$$f_G(x) = per[xI - L(G)] = \sum_{t=0}^{n} (-1)^t a_t(G) x^{n-t}.$$

(13)

Since the permanent is invariant under permutation similarities, the coefficients and roots of $f_G(x)$ are graph-theoretic invariants. The following result is the permanental analog of Theorem 2.9.

Theorem 6.1 [36]: The multiplicity of 1 as a root of the permanent& polynomial $f_G(x)$ is at least $p(G) - q(G)$.

There are, of course, a number of obstacles to be overcome in the study of $f_G(x)$, not the least of which is the notorious computational intractability of the permanent function. (Only in some special cases, e.g. for trees, has this challenge been overcome [13, 83].) Another obstacle concerns the roots of $f(x)$.

For one thing, there is nothing resembling eigenvectors associated with them. For another, they need not all be real. [If, for example, A is an n-by-n correlation matrix, then the roots of per (xI-A) are all real if and only if $A = I_n$.] Say that a graph is permanently real if all roots of $f_G(x)$ are real. Then there is not a single permanently real graph among the 112 connected graphs on 6 vertices. But if $G = K_{1,n-1}$, then (Theorem 6.1) $(X - I)^{n-2}$ is a factor of $f_G(x)$. The other factor is $x^2 - nx + 2(n - 1)$. Thus, $K_{1,n-1}$, is permanently real for all $n \geq 7$.

Denote the characteristic polynomial of $L(G)$ by

$$\det[xI - L(G)] = \sum_{t=0}^{n-1}(-1)^t b_t(G)x^{n-t}.$$

$$(14)$$

Graph-theoretic interpretations of the coefficients $b_t(G)$ were given in [30, 75, 761. (See [45] for the edge-weighted version.)

Theorem 6.2: If G is a graph on n vertices, then

$$b_t(G) = \sum P(F),$$

where the sum is over all (n - t)-edged spanning forests F of G, and P(F) is the product of the numbers of vertices in each of the t components of F.

As in (10), denote the chromatic polynomial of G by

$$p_G(x) = \sum_{t=0}^{n-1}(-1)^t c_t(G)x^{n-t}.$$

Theorem 6.3: Let G be a connected graph on n vertices. Then $a_0(G) = b_0(G) = c_0(G) = 1$, $a_1(G) = b_1(G) = 2c_1(G) = 2m$, and

$$a_t(G) \geq b_t(G) \geq (t+1)c_t(G), \quad 1 < t < n.$$
(15)

In (151, the left-hand inequality is immediate from Schur's theorem [126]. In the right-hand inequality [93], equality holds for $t = n - 1$ if and only if G is a tree; if $n \geq 4$, then equality holds for $t = n - 2$ if and only if it holds for all t if and only if $G = K_{1,n-1}$.

Now, determinants and permanents are but two examples of matrix functions that have come to be known as immanants. If x is an irreducible (characteristic 0) character of S_n the corresponding immanant, d_x is defined by

$$d_x(A) = \sum_{p \in S_n} X(p) \prod_{t=1}^{n} a_{tp(t)}$$
(16)

for any n-by-n matrix $A = (a_{ij})$. If $x = \varepsilon$, the signurn character, then $d_{x'} = \det$. If $x = 1$, then $d_x = \text{per}$.

There is a natural one-to-one correspondence between the irreducible characters x of S_n and the (nonincreasing, integer) partitions of n. (Thus, Ferrers-Sylvester diagrams, such as those in Figure 2, play a prominent role in the character theory of S_n) Those characters x_r corresponding to partitions of the form $(r, 1^{n-r})$, short for

$$n = r + 1 + \cdots + 1,$$

are called single-hook characters. For example, $x_1 = \varepsilon$ and $x_n = 1$. [In general, $x_r(e) = C(n - 1, r - 1)$, the binomial coefficient (n - l)-choose- (r – 1).] We will denote by d_r the immanant corresponding to x_r, so $d_1 = \det$, and $d_{n'} = \text{per}$. [The context should permit the reader to distinguish between the immanant $d_r(L(G))$, the determinantal divisor $d_r(G)$, and the vertex degree d_r.] It turns out that these single-hook immanants can be used to count h(G), the number of Hamiltonian circuits in G [91]. (Also, see [53].)

Theorem 6.4: Let G be a connected graph on n $\geq$ 3 vertices. Then the number of Hamiltonian circuits in G is

$$h(G) = \frac{1}{2n} \sum_{r=2}^{n} (-1)^r d_r(L(G)).$$

If A is an n-by-n positive semidefinite Hermitian matrix, then (n - l)'per A $\geq$ $d_{n-1}(A)$ [99]. 0 nce again, it would seem surprising if a general result like this could not be improved when restricted to a much smaller class of matrices.

Conjecture 6.5: If T is a tree on n vertices, then (n - 2) per L(T) $\geq$ $d_{n-1}(L(T))$.

There is a natural affinity between immanants and Laplacians. Recall that L(G) is a Gram matrix based, e.g., on the row vectors Q_1, Q_2, . . . , Q_m of Q(G). It turns out that $d_x(L(G))$ is n!/x(e) times the length of the "decomposable symmetrized tensor" $Q_1 * Q_2 * ** . * Q_m$. [If G is connected, then $d_x(L(G)) > 0$ for all x $\neq$ ε.]

Let $M_x = (U: d_x(U^{-1}AU) = d_x(A)$ for all A}. If x = ε, then M_x is the full linear group. Otherwise, it is the monomial group consisting of all nonzero scalar multiples of permutation matrices [48]. Thus, each of the immanantal polynomials $d_x(XI - L(G))$, as x ranges over the characters of S_n (irreducible or not), is a graph-theoretic invariant.

At present, only a little is known about general immanantal roots. While they need not all be real, those that are lie in the interval [0, $\lambda_1(G)$] [89]. After determinant and permanent, the most widely studied immanantal polynomial is $d_2(xI - L(G))$. It is known that the d_2-roots lie in the Gerigorin circles [72]. The coefficient of x: in the d_2-polynomial is related to moment sums in graphs, leading to an extension of the notion of centroid point [92].

It turns out that immanantal polynomials, even when they are all taken together, are not much better than the characteristic polynomial when it comes to distinguishing nonisomorphic graphs. J. Turner [135] found a pair of nonisomorphic trees T and T' on 12 vertices such that $d_x(xI - A(T')) = d_x(xI - A(T'))$ for all 77 irreducible characters x of S_{12}. Such examples also exist for the Laplacian. Indeed, as we now see, they are typical (in the sense of Schwenk and McKay).

Theorem 6.6 [12]: Let t_n be the number of nonisomorphic trees on n vertices. Let r_n be the number of such trees T for which there is a nonisomorphic tree T' such that, simultaneously, for every character x of S_n both

$d_x(xI - A(T)) = d_x(xI - A(T'))$ and

$d_x(xI - L(T)) = d_x(xI - L(T'))$

Then lim $_{n \to \infty} r_n / t_n = 1$.

We have used (iv) and (v) in the statement of Theorem 6.6 because it may be viewed as a continuation of Theorem 5.2.

REFERENCES

1. J. Aihara, Why aromatic compounds are stable, Sci. Amer. 266, Mar. 1992, pp. 62-68.
2. N. Alon, Eigenvalues and expanders, Combinatorics 6:83-96 (1986).
3. N. Alon, Z. Gahl, and V. D. Milman, Better expanders and superconcentrators, 1. Algorithms 8:337-347 (1987).
4. N. Alon and V. D. Milman, Ai, isoperimetric inequalities for graphs, and superconcentrators,]. Combin. Theory Ser. B 38:73-88 (1985).
5. W. N. Anderson and T. D. Morley, Eigenvahres of the Laplacian of a graph, Linear and M&linear Algebra 18:141-145 (1985).
6. A. T. Balaban, Chemical graphs, Theoret. Chim. Acta (BerZ.) 53:355-375 (1979).
7. R. B. Bapat, A bound for the permanent of the Laplacian matrix, Linear Algebra AppZ. 74:219-223 (1986).
8. R. B. Bapat and G. Constantine, An enumerating function for spanning forests and color restrictions, Linear Algebra A?pZ. 173:231-237 (1992).

9. K. A. Berman, Bicycles and spanning trees, SIAM]. Algebraic Discrete Methods 7:1-12 (1986).

10. F. Bien, Constructions of telephone networks by group representations, Notices Amer. Math. Sot. 36:5-22 (1989).

11. N. Biggs, Algebraic Graph Theory, Cambridge U.P., 1974.

12. P. Botti and R. Merris, Almost all trees share a complete set of immanantal polynomials, J. Graph Theory, to appear.

13. P. Botti, R. Merris, and C. Vega, Laplacian permanents of trees, SIAM J. Discrete Math. 5:460-466 (1992).

14. R. A. Brualdi and J. L. Goldwasser, Permanent of the Laplacian matrix of trees and bipartite graphs, Discrete Math. 48:1-21 (1984).

15. F. C. Bussemaker and D. M. Cvetkovib, There are exactly 13 connected, cubic, integral graphs, Univ. Beograd PubZ. Elektrotehn. Fak. Ser. Mat. Fiz. 544-576:43-48 (1976).

16. S. Chaiken, A combinatorial proof of the all minors matrix tree theorem, SIAM J. Algebraic Discrete Methods 3:319-329 (1982).

17. S. Chaiken and D. J. Kleitman, Matrix tree theorems, J. Combin. Theory Ser. A 24:377-381 (1978).

18. J. Cheeger, A lower bound for the smallest eigenvalue of the Laplacian, in Problems in Analysis (R. C. Gunning, Ed.), Princeton U.P., 1970, pp. 195-199.

19. F. R. K. Chung, Diameters and eigenvalues, J. Amer. Math. Sot. 2:187-196 (1989).

20. F. R. K. Chung, V. Faber, and T. Manteuffel, An upper bound on the diameter of a graph from eigenvalues associated with its Laplacian, manuscript.

21. K. L. Collins, On a conjecture of Graham and Lovasz about distance matrices, Discrete AppZ. Math. 25:27-35 (1989).

22. Factoring distance matrix polynomials, Discrete Math., to appear.

23. Private communication.

24. G. M. Constantine, Schur convex functions on the spectra of graphs, Discrete Math. 45:181-188 (1983).

25. Combinatorial Theory and Statistical Design, Wiley, New York, 1987.

26. Graph complexity and the Laplacian matrix in blocked experiments, Linear and Multilinear Algebra 28:49-56 (1990).

27. D. M. Cvetkovih, Cubic integral graphs, Univ. Beograd PubZ. Elektrotehn. Fak. Ser. Mat. Fiz. 498-541:107-113 (1975).

28. D. M. Cvetkovi& and M. Doob, Developments in the theory of graph spectra, Linear and Multilinear AZgebra 18:153-181.

29. D. M. Cvetkovib, M. Doob, I. Gutman, and A. Torgasev, Recent Results in the Theory of Graph Spectra, North-Holland, Amsterdam, 1988.

30. D. M. CvetkoviE, M. Doob, and H. Sachs, Spectra of Graphs, Academic, New York, 1979.

31. E. A. Dinic, A. K. Kelmans, and M. A. Zaitsev, Nonisomorphic trees with the same T-polynomial, Inform. Process. Lett. 6:73-76 (1977).

32. M. Edelberg, M. R. Garey, and R. L. Graham, On the distance matrix of a tree, Discrete Math. 14:23-31 (1976).

33. B. E. Eichinger, Elasticity theory. I. Distribution functions for perfect phantom networks, Macromolecules 5:496-505 (1972).

34. Configuration statistics of Gaussian molecules, Macromolecules, 13:1-11 (1980).

35. Random elastic networks. I. Computer simulation of linked stars, J, Chem. Phys. 75:1964-1979 (1981).

36. I. Faria, Permanental roots and the star degree of a graph, Linear Algebra Appl. 64:255-265 (1985).

37. M. Fiedler, Algebraic connectivity of graphs, Czechoslovak Math. J, 23:298-305 (1973).

38. Eigenvectors of acyclic matrices, Czechoslovak Math. J. 25:607-618 (1975).

39. A property of eigenvectors of nonnegative symmetric matrices and its application to graph theory, Czechoslovak Math. 1. 25:619-633 (1975).

40. Algebraische Zusammenhangszahl der Graphen und ihre Numerische Bedeutung, Internat. Ser. Numer. Math. 29, Birkhzuser, Basel, 1975, pp. 69-85.

41. An algebraic approach to connectivity of graphs, in Recent Advances in Graph Theory, Academia, Prague, 1975, pp. 193-196.

42. Laplacian of graphs and algebraic connectivity, in Combinatorics and Graph Theory, Banach Center Publ. 25, PWN, Warsaw, 1989, pp. 57-70.

43. Absolute algebraic connectivity of trees, Linear and M&linear Algebra 26:85-106 (1990).

44. A geometric approach to the Laplacian matrix of a graph, presented at Algebraic Graph Theory Workshop, Inst. for Math. Appl., Minneapolis, 12 November 1991.

45. M. Fiedler and J. Sedllfek, 0 W-basich orientovanjch grafti (in Czech), das. P&t Mat. 83:214-225 (1958).

46. M. E. Fischer, On hearing the shape of a drum, J. Combin. Theory 1:105-125 (1966).

47. W. C. Forsman, Graph theory and the statistics and dynamics of polymer chains, J. Chem. Phys. 65~4111-4115 (1976).

48. S. Friedland, Maximality of the monomial group, Linear and Mu&linear Algebra 18:1-7 (1985).

49. Quadratic forms and the graph isomorphism problem, Linear Algebra Appl. 150:423-442 (1991).

50. Lower bounds for the first eigenvalue of certain M-matrices associated with graphs, Linear Algebra Appl. 172:71-84 (1992).

51. C. Godsil, Real graph polynomials, in Progress in Graph Theory (J. A. Bondy and U. S. R. Murty, Eds.), Academic, New York, 1984, pp. 281-293.

52. J. L. Goldwasser, Permanent of the Laplacian matrix of trees with a given matching, Discrete Math. 61:197-212 (1986).

53. I. P. Goulden and D. M. Jackson, The enumeration of directed closed Euler trails and directed Hamiltonian circuits by Lagrangian methods, European J. Combin. 2:131-135 (1981).

54. R. L. Graham and L. Lo&z, Distance matrix polynomials of trees, in Theory and Applications of Graphs (Y. Alavi and D. R. Lick, Eds.), Lecture Notes in Math. 642, Springer-Verlag, Berlin, 1978, pp. 186-190.

55. Distance matrix polynomials of trees, Adv. in Math. 29:60-88 (1978).

56. R. L. Graham and H. 0. Pollak, On the addressing problem for loop switching, Bell System Tech. J. 50:2495-2519 (1971).

57. M. Gromov and V. D. Milman, A topological application of the isoperimetric inequality, Amer. J. Math. 105:843-854 (1983).

58. R. Crone and R. Merris, Algebraic connectivity of trees, Czechoslovuk Math. J. 37:660-670 (1987).

59. Cutpoints, lobes and the spectra of graphs, Portugal. Math. 45:181-188 (1988).

60. Ordering trees by algebraic connectivity, Graphs Combin. 6:229-237 (1990).

61. Coalescence, majorization, edge valuations and the Laplacian spectra of graphs, Linear and Multilinear Algebra 27:139-146 (1990).

62. The Laplacian spectrum of a graph II, SIAM J. Discrete Math., to appear.

63. R. Grone, R. Merris, and V. S. Sunder, The Laplacian spectrum of a graph, SIAM J. Matrix Anal. AppZ. 11:218-238 (1990).

64. R. Crone, R. Merris, and W. Watkins, Laplacian unimodular equivalence of graphs, in Combinatorial and Graph-Theoretic Problems in Linear Algebra (R. Brualdi, S. Friedland, and V. Klee, Eds.), IMA Vol. Math. AppZ. 50, SpringerVerlag, New York, to appear.

65. R. Crone and G. Zimmermann, Large eigenvalues of the Laplacian, Linear and Multilinear Algebra 28~45-47 (1990).

66. I. Gutman, Graph-theoretical formulation of Forsman's equations, J. Chem. Phys. 68:1321-1322 (1978).

67. I. Gutman and D. H. Rouvray, A new theorem for the Wiener molecular branching index of trees with perfect matchings, Comput. and Chem., to appear.

68. F. Harary and A. J. Schwenk, Which graphs have integral spectra?, in Graphs and Combinatorics (R. A. Bari and F. Harary, Eds.), Lecture Notes in Math. 406, Springer-Verlag, Berlin, 1974, pp. 45-51.

69. Y. Hattori, Nonisomorphic graphs with the same T-polynomial, Infinn. Process. Lett. 22:133-134 (1986).

70. H. Hosoya, Topological index. A newly proposed quantity characterizing the topological nature of structural isomers of saturated hydrocarbons, Bull. Chem. Sot. Jupun 44:2332-2339 (1971).

71. H. Hosoya, M. Murakami, and M. Gotoh, Distance polynomial and characterization of a graph, Nut. Sci. Rep. Ochunomizu Univ. 24~27-34 (1973).

72. C. R. Johnson, R. Merris, and S. Pierce, Inequalities involving immanants and diagonal products for H-matrices, Portugal. Math. 43:43-54 (1985-1986).

73. F. Juhasz, The asymptotic behavior of Fiedler's algebraic connectivity of graphs, Discrete Math. 96:59-63 (19911.

74. M. Kac, Can one hear the shape of a drum, Amer. Math. Monthly 73:1-23 (1966).

75. A. K. Kel'mans, Properties of the characteristic polynomial of a graph, (in Russian) Kibernet. na SluZbu Kommunizmu 4:27-41 (1967).

76. A. K. Kelmans and V. M. Chelnokov, A certain polynomial of a graph and graphs with an extremal number of trees, J. Combin. Theory Ser. I3 16:197-214 (1974).

77. G. Kirchhoff, cber die Au&sung der Gleichungen, auf welche man bei der Untersuchung der linearen Verteilung galvanischer Strijme gefuhrt wird, Ann. Phys. Chem. 72:497-508 (1847).

78. M. Lewin, A generalization of the matrix-tree theorem, Math. Z. 181:55-70 (1982).

79. D. J. Lorenzini, Arithmetical graphs, Math. Ann. 285:481-501 (1989).

80. A finite group attached to the laplacian of a graph, Discrete Math. 91:277-282 (1991).

81. L. Lo&z, On the Shannon capacity of a graph, IEEE Trans. Inform. Theory 25:1-7 (1979).

82. C. Maas, Transportation in graphs and the admittance spectrum, Discrete Appl. Math. 16:31-49 (1987).

83. M. Malek, Conversion of the permanent into the determinant, manuscript.

84. A. W. Marshall and I. Olkin, Inequalities: Theory of Majorization and Its Applications, Academic, New York, 1979.

85. S. B. Maurer, Matrix generalizations of some theorems on trees, cycles and cocycles in graphs, SIAM J. Appl. Math. 30:143-148 (1976).

86. B. D. McKay, On the spectral characterisation of trees, Ars Combin. 3:219-232 (1977).

87. The expected eigenvalue distribution of a large regular graph, Linear Algebra Appl. 40:203-216 (1981).

88. Private communication.

89. R. Merris, Two problems involving Schur functions, Linear Algebra Appl. 10:155-162 (1975).

90. The Laplacian permanental polynomial for trees, Czechoslovak Math. 1. 32:397-403 (1982).

91. Single-hook characters and Hamiltonian circuits, Linear and Multilinear Algebra 4:21-35 (1983).

92. The second immanantal polynomial and the centroid of a graph, SIAM J. Algebraic Discrete Methods 7:484-503 (1986).

93. Characteristic vertices of trees, Linear and M&linear Algebra 22:115-131 (1987).

94. An edge version of the matrix-tree theorem and the Wiener index, Linear and Multilinear Algebra 25:291-296 (1989).

95. The distance spectrum of a tree, J. Graph Theory 14:365-369 (19901.

96. The number of eigenvalues greater than two in the Laplacian spectrum of a graph, Portugal. Math. 48:345-349 (1991).

97. Unimodular equivalence of graphs, Linear Algebra Appl. 173:181-189 (1992).

98. R. Merris, K. R. Rebman, and W. Watkins, Permanental polynomials of graphs, Linear Algebra Appl. 38:273-288 (1981).

99. R. Merris and W. Watkins, Inequalities and identities for generalized matrices, Linear Algebra Appl. 64~223-242 (1985).

100. M. Michalski, Branching extent and spectra of trees, Publ. Inst. Math. 39:35-43 (1986).

101. B. Mohar, Laplacian matrices of graphs, Preprint Ser. Dept. Math. Univ. E. K. Ljubljana 26:385-392 (1988).

102. Isoperimetric inequalities, growth, and the spectra of graphs, Linear AZgebru AppZ. 103:119-131 (1988).

103. Isoperimetric numbers of graphs, J. Combin. Theory Ser. B 47:274-291 (1989).

104. The Laplacian spectrum of graphs, in Graph Theory, Combinatorics, and A$-cations (Y. Alavi, G. Chartrand, O. R. Ollermann, and A. J. Schwenk, Eds.), Wiley, New York, 1991, pp. 871-898.

105. Eigenvalues, diameter, and mean distance in graphs, Graphs Combin. 7:53-64 (1991).

106. B. Mohar and T. Pisanski, How to compute the Wiener index of a graph, J. Math. Chem. 2~267-277 (1988).

107. B. Mohar and S. Poljak, Eigenvalues and the max-cut problem, CzechosZowak Math. J. 40:343-352 (1990).

108. Eigenvalues in combinatorial optimization, IMA Preprint Ser. 939, Inst. for Math. Appl., Univ. of Minnesota, Minneapolis, 1992.

109. I. Motoc and A. T. Balaban, Topological indices: Intercorrelations, physical meaning, correlational ability, Reu. Roumuine Chim. 26:593-600 (1981).

110. N. A. Neuburger and B. E. Eichinger, Computer simulation of the dynamics of trifunctional and tetrafunctional end-linked random elastic networks, J. Chem. Phys. 83:884-891 (1985).

111. A. Nilli, On the second eigenvalue of a graph, Discrete Math. 91:207-210 (1991).

112. S. Pierce, Permanents of correlation matrices, in Current Trends in Matrix Theory (R. Grone et al., Eds.), Elsevier, Amsterdam, 1987, pp. 247-249.

113. O. E. Polansky and D. Bonchev, Theory of the Wiener number of a graph. II. Transfer graphs and some of their metric properties, Match 25:3-39 (1990).

114. A. Pothen, H. D. Simon, and K.-P. Liou, Partitioning sparse matrices with eigenvectors of graphs, SIAM J. Matrix Anal. AppZ. 11:430-452 (1990).

115. M. Protter, Can one hear the shape of a drum? revisited, SIAM Rev. 29:185-197 (1987).

116. R. C. Read, An introduction to chromatic polynomials, J. Con&n. Theory 4:52-71 (1968).

117. D. H. Rouvray, Should we have designs on topological indices, in Chemical Applications of TopoZogy and Graph Theory (R. B. King, Ed.), Elsevier, Amsterdam, 1983, pp. 159-177.

118. Predicting chemistry from topology, Sci. Amer. 255, Sept. 1986, pp. 40-47.

119. The role of the topological distance matrix in chemistry, in Mathematics and Computational Concepts in Chemistry (N. Trinajstib, Ed.), Horwood, Chichester, 1986, pp. 295-306.

120. Characterization of molecular branching using topological indices, in AppZications of Discrete Mathematics (R. 0. Ringesen et al., Eds.), SIAM, Philadelphia, 1988, pp. 176-178.

121. The challenge of characterizing branching in molecular species, Discrete AppZ. Math. 19:317-338 (1988).

122. E. Ruth and I. Gutman, The branching extent of graphs, J. Combin. Inform. System Sci. 4:285-295 (1979).

123. S. N. Ruzieh and D. L. Powers, The distance spectrum of the path I',, and the first distance eigenvector of connected graphs, Linear and Multilinear Algebra 28:75-81 (1990).

124. E. M. de Sa, Note on graphs and weakly cyclic matrices, Discrete Math. 34:275-281 (1981).

125. I. Schur, Uber eine Klasse von Mittelbildungen mit Anwendungen die Determinanten, Sitzungsber. Berlin. Math. Gesellschafi 22:9-20 (1923).

126. Uber endliche Gruppen und Hermitesche Formen, Math. Z. 1:184-207 (1918).

127. S. Schuster, Interpolating theorem for the number of end-vertices of spanning trees, J. Graph Theory 7:203-208 (1983).

128. A. J. Schwenk, Almost all trees are cospectral, in New Directions in Graph Theory (F. Harary, Ed.), Academic, New York, 1973, pp. 275-307.

129. Exactly thirteen connected cubic graphs have integral spectra, in Theory and Applications of Graphs (Y. Alavi and D. R. Lick, Eds.), Lecture Notes in Math. 642, Springer-Verlag, Berlin, 1978, pp. 516-553.

130. L. Y. Shy and B. E. Eichinger, Large computer simulations on elastic networks: Small eigenvalues and eigenvalue spectra of the Kirchhoff matrix, 1. Chem. Phys. 90:5179-5189 (1989).

131. W. So, Rank one perturbation and its application to the Laplacian spectrum of a graph, manuscript.

132. H. M. Trent, A note on the enumeration and listing of all maximal trees of a connected linear graph, Proc. Nat. Acad. Sci. U.S. A. 40:1004-1007 (1954).

133. N. Trinajstib, Chemical Graph Theory, Vol. I, CRC Press, Boca Raton, Fla., 1983.

134. Chemical Graph Theory, Vol. II, CRC Press, Boca Raton, Fla., 1983.

135. J. Turner, Generalized matrix functions and the graph isomorphism problem, SIAM J. AppZ. Math. 16:520-536 (1968).

136. Y. C. de Verdi&e, Sur un nouvel invariant des graphes et un critere de planarit&, J. Combin. Theory Ser. B 50:11-21 (1990).

137. Spectres de variktes Riemanniennes et spectres de graphes, manuscript.

138. A. Vrba, The permanent of the Laplacian matrix of a bipartite graph, Czechoslovak Math. J. 36:397-403 (1982).

139. Principal subpermanents of the Laplacian matrix, Linear and Multilinear Algebra 19:335-346 (1986).

140. W. Watkins, The Laplacian matrix of a graph: Unimodular congruence, Linear and M&linear Algebra 28:35-43 (1990).
141. Unimodular congruence of the Laplacian matrix of a graph, Linear Algebra Appl., to appear.
142. D. Welsh, Matroids and their applications, in Selected Topics in Graph Theory 3 (L. W. Beineke, and R. J. Wilson, Eds.), Academic, New York, 1988, pp. 43-70.
143. H. Whitney, A logical expansion in mathematics, Bull. Amer. Math. Sot. 38:572-579 (1932).
144. 2-isomorphic graphs, Amer. J. Math. 55:245-254 (1933).
145. P. J. Cameron, J. M. Goethals, J. J. Seidel, and E. E. Shult, Line graphs, root systems, and elliptic geometry, J. Algebra 43:305-327 (1976).
146. B. Cipra, You can't hear the shape of a drum, Science 255:1642-1643 (27 Mar. 1992).
147. C. Godsil, D. A. Holton, and B. McKay, The spectrum of a graph, in Corn&&orial Mathematics V, Lecture Notes in Math. 622, Springer-Verlag, Berlin, 1977, pp. 91-117.
148. C. Gordon, D. L. Webb, and S. Wolpert, One cannot hear the shape of a drum, Bull. Amer. Math. Sot. 27:134-138 (1992).
149. I. P. Goulden and D. M. Jackson, Immanants of combinatorial matrices, J. Algebra 148:305-324 (1992).
150. M. Fiedler, Aggregation in graphs, Colloq. Math. Sot. Jinos Bolyai 18:315-330 (1976).
151. An estimate for the nonstatistic eigenvalues of doubly stochastic matrices, IMA Preprint Ser. 902, Inst. for Math. Appl., Univ. of Minnesota, 1991.
152. J. J. Seidel, Distance matrices and Lorentz space, in Workshop VZudimir USSR, Aug. 1991, pp. l-2.
153. J. A. Sjogren, Cycles and spanning trees, Math. Comput. Modelling 15:87-102 (1991).
154. I. Faria, Multiplicity of integer roots of polynomials of graphs, Linear Algebra Appl., to appear.
155. R. Grone, On the geometry and Laplacian of a graph, Linear Algebra Appl. 150:167-178 (1991).

CITATION

Russell Merris, Laplacian matrices of graphs: a survey, Linear Algebra and its Applications, Volumes 197–198, January–February 1994, Pages 143-176, ISSN 0024-3795, http://dx.doi.org/10.1016/0024-3795(94)90486-3.

A New Expression for Rhotrix

Abdul Mohammed
Department of Mathematics, Ahmadu Bello University,
Zaria, Nigeria

ABSTRACT

This paper presents a new technique for expressing rhotrices in a generalize form. The method involves using multiple array indexes as analogous to matrix expressions, unlike the earlier method in the literature, which can only be functional in a single array computational environment. The new rhotrix look will encourage the study of rhotrix algebra and analysis from a better perspective. In addition, computing efficiency and accuracy will also be improved, particularly when the operations in rhotrix space over the new expression are algorithmatized for computing machines.

INTRODUCTION

Rhotrix theory is a relatively new area of Mathematics, whose goal is central on representing arrays of numbers in rhomboid mathematical form, unlike matrix theory dealing with representing array of numbers in rectangular form. The concept of rhotrices of size three, also well known as base rhotrices, was introduced by Ajibade [1] as an extension of ideas on matrix-tertion and matrix-noitret, suggested by Atanassov and Shannon [2] .

Expressing rhotrices of size n, particularly in a generalized form, has been a difficult problem to the rhotrix theorist. This may probably be the reason, why most works in rhotrix theory are communicated using rhotrices of specific size; one can see [1] [3] - [6]. Though, some attempts were made by Sani [8] and Mohammed et al. [7] to overcome such problem in literature, but the two generalizations proposed by these authors are sometimes inconvenient for presenting rhotrices as two dimensional objects.

Thus, it becomes imperative to seek for a new method of generalizing expression for rhotrices in order to allow for exactness, efficiency and convenience in presentation of research results. Furthermore, better algebra and analysis of rhotrix theory can be studied using multiple array technique as proposed in this article.

AN OVERVIEW OF INITIAL GENERALIZED RHOTRIX EXPRESSIONS

Coupled Matrix Technique for Expressing Rhotrix in a Generalized Form

In an attempt to find an answer to the problem of "finding a transformation of rhotrix to matrix and vice versa" posed by Ajibade [1] in the concluding remark of his article, Sani [8] proposed an alternative method for multiplication of rhotrices of size 3, based on their rows and columns vectors, as comparable to matrices, recorded as follows:

$$R(3) \circ Q(3) = \left\langle \begin{matrix} & a & \\ b & h(R) & d \\ & e & \end{matrix} \right\rangle \circ \left\langle \begin{matrix} & f & \\ g & h(Q) & j \\ & k & \end{matrix} \right\rangle =$$

$$\left\langle \begin{matrix} & af + dg & \\ bf + eg & h(R)h(Q) & aj + dk \\ & bj + ek & \end{matrix} \right\rangle$$

This alternative multiplication approach was also used in his work to establish some relationships between rhotrices of size 3 and 2 × 2 dimensional matrices through an isomorphism. Thereafter, Sani [9] extend his own method for multiplication of base rhotrices to higher size rhotrices in form of generalization, recorded as follows:

$$R(n) \circ Q(n) = \left\langle a_{i1j1}, c_{l1k1} \right\rangle \quad \circ \left\langle b_{i2j2}, d_{l2k2} \right\rangle = \left\langle \sum_{i2j1=1}^{m} (a_{i1j1} b_{i2j2}), \sum_{i2k1=1}^{m-1} (c_{l1k1}, d_{l2k2}) \right\rangle,$$

where

$$R(n) = \left\langle \begin{array}{ccccccc} & & & a_{11} & & & \\ & & a_{21} & c_{11} & a_{12} & & \\ & \cdots & \cdots & \cdots & \cdots & \cdots & \\ a_{m1} & \cdots & \cdots & \cdots & \cdots & \cdots & a_{mm} \\ & \cdots & \cdots & \cdots & \cdots & \cdots & \\ & & a_{mm-1} & c_{m-1m-1} & a_{m-1m} & & \\ & & & a_{mm} & & & \end{array} \right\rangle,$$

a_{ij} and a_{kl} represent the a_{ij} and C_{kl} elements respectively, m=(n+1)/2 with i, j=1, 2, 3, L, m and k, l=1, 2, 3, L, m-1 This idea of expressing rhotrix of size n, n 2Z⁺+1, as a combination of two matrices, was later presented as an idea for conversion of a rhotrix to a special form of a matrix, termed "coupled matrix" in Sani [10] . That is, a rhotrix R of size n can be expressed as a couple of two matrices A and C of sizes m × m and (m-1) × (m-1) respectively, where m=(n+1)/2 and n 2Z⁺+1.

It is noteworthy to mention that, this method presented by Sani for expressing rhotrix in a generalised form, requires some form of transformation or half transpose, in order to identify the rows and columns for any given rhotrix. Also, unique expression for the "rhotrix heart" cannot be deduced and therefore, this method of rhotrix expression is unsuitable for presenting heart-based rhotrices.

Single Array Technique for Expressing Rhotrix in a Generalized Form

Mohammed et al. [7] employ a single dimensional array routine for expressing real rhotrices of size n in a generalised form, recorded as follows:

$$
R(n) = \left\langle \begin{array}{ccccccccc}
 & & & & r_1 & & & & \\
 & & & r_2 & r_3 & r_4 & & & \\
 & & \vdots & \vdots & \vdots & \vdots & \vdots & & \\
 & \vdots & \vdots & \vdots & \vdots & \vdots & \vdots & \vdots & \\
r_{\left\{\frac{t+1}{2}\right\}-n\backslash 2} & \cdots & \cdots & \cdots & r_{\left\{\frac{t+1}{2}\right\}} & \cdots & \cdots & \cdots & r_{\left\{\frac{t+1}{2}\right\}+n\backslash 2} \\
 & \vdots & \vdots & \vdots & \vdots & \vdots & \vdots & \vdots & \\
 & & \vdots & \vdots & \vdots & \vdots & \vdots & & \\
 & & & r_{t-3} & r_{t-2} & r_{t-1} & & & \\
 & & & & r_t & & & &
\end{array} \right\rangle ,
$$

where

$$
t = \frac{1}{2}\left(n^2 + 1\right),\ n \in 2Z^+ + 1
$$

and n/2 is the integer value from division of n by 2.

The entry $h(R) = r_{\frac{t+1}{2}}$ is called the heart of any rhotrix R(n). The operations of addition, scalar multiplication and multiplication (o) with respect to this generalization were discussed in [7].

At this point, it is necessary to point out that this method of expressing rhotrices in generalized form follow the initial definition given for rhotrices in [1] but it can only be functional on computing machines using single array procedure, which is inadequate and capable of causing slow processing of operations involving rhotrices having the same bigger size. Also, entries in rhotrices cannot be immediately indicated.

A NEW GENERALIZED EXPRESSION FOR RHOTRIX

Since rhotrix theory is a relatively new paradigm of matrix theory, finding a generalized expression that precisely represents rhotrices of the same size n, while preserving row and column entries, cannot be divorced from multiple array indexes as obtainable in matrices. For this reason, an adoption of the procedure that is analogous to representation of matrices will be utilized.

For example, consider a zero heart rhotrix A of size n = 5 given below;

$$A(5) = \left\langle \begin{array}{ccccc} & & 1 & & \\ & 4 & 3 & 8 & \\ 10 & -7 & 0 & -6 & 5 \\ & -4 & 7 & 9 & \\ & & -8 & & \end{array} \right\rangle.$$

Clearly, if we use comparison with matrix row and column entries then by observation, the entry value 1 in rhotrix A can be found at row 1 and column 3; the entry value 8 in rhotrix A can be found at row 2 and column 4; the entry value-4 in rhotrix A can be located at row 4 and column 2; the entry value 0 in rhotrix A can be located at row 3 and column 3; and the entry value 8 in rhotrix A can be found at row 5 and column 3. Thus, we can adopt this procedure to give a new expression for the rhotrix R(n).

Let $\overset{\diamond}{C}(n)$ be the set of all rhotrices of size n over a field F (real or complex). Since any rhotrix $B(n) \in \overset{\diamond}{C}$ has total number of entries as $\frac{1}{2}(n^2 + 1)$, where n $2Z^+$+1 then a technique for expressing the set of all rhotrices of the same size n in a generalized form is given by

$$
\overset{\lozenge}{C}(n) = \left\{ \left\langle \begin{array}{ccccccc} & & & & c_{1,\frac{n+1}{2}} & & \\ & & & c_{2,\frac{n-1}{2}} & c_{2,\frac{n+1}{2}} & c_{2,\frac{n+3}{2}} & \\ & \vdots & \vdots & \vdots & \vdots & \vdots & \vdots \\ c_{\frac{n+1}{2},1} & c_{\frac{n+1}{2},2} & \cdots & \cdots & c_{\frac{n+1}{2},\frac{n+1}{2}} & \cdots & \cdots & c_{\frac{n+1}{2},n-1} & c_{\frac{n+1}{2},n} \\ & \vdots & \vdots & \vdots & \vdots & \vdots & \vdots \\ & & & c_{n-1,\frac{n-1}{2}} & c_{n-1,\frac{n+1}{2}} & c_{n-1,\frac{n+3}{2}} & \\ & & & & c_{n,\frac{n+1}{2}} & & \end{array} \right\rangle : c_{i,j} \in F \right\}
$$

(1)

where, c_{ij} are entries from a field of real or complex numbers F. It is clear that any entry b_{ij} in B(n) must be indicated by i^{th} row and j^{th} column. This method of expressing rhotrices in a general form is analogous to that of matrices because any entry c_{ij} within a rhotrix is indicated out by its row i and column j, which is analogous to any entry x_{ij} in matrix set X of all m × n dimensional matrices, denoted by

$$
(n,m) = \left\{ \begin{bmatrix} x_{11} & x_{12} & \cdots & x_{1n} \\ x_{21} & x_{22} & \cdots & x_{2n} \\ \vdots & \vdots & \ddots & \vdots \\ x_{m1} & x_{m2} & \cdots & x_{mn} \end{bmatrix} : x_{11}, x_{12}, \cdots, x_{mn} \in F \right\}
$$

From Equation (1), $h(C) = c_{\frac{n+1}{2},\frac{n+1}{2}}$ is called the heart of any rhotrix C(n) $\in$ C(n).

Thus, for n = 3, we get

$$
C(3) = \left\langle \begin{array}{ccc} & c_{11} & \\ c_{21} & c_{22} & c_{23} \\ & c_{32} & \end{array} \right\rangle
$$

and for n = 5, we get

$$C(5) = \left\langle \begin{array}{ccccc} & & c_{13} & & \\ & c_{22} & c_{23} & c_{24} & \\ c_{31} & c_{32} & c_{33} & c_{34} & c_{35} \\ & c_{42} & c_{43} & c_{44} & \\ & & c_{53} & & \end{array} \right\rangle .$$

Operations in Rhotrix Space over the New Expression

Let R and S be two rhotrices of the same size n and let a be a scalar. If we denote the entries in rhotrix R as r_{ij} and the entries in rhotrix S as S_{ij}, then we can extend the operations of addition, scalar multiplication and multiplication (o) defined for heart-based rhotrices of size-3 in [1] to rhotrices of the same n-size, using the rhotrix new expression respectively as follows:

$$R + S = \left\langle \begin{array}{ccccc} & & & & r_{1,\frac{n+1}{2}} + S_{1,\frac{n+1}{2}} \\ & & r_{2,\frac{n-1}{2}} + S_{2,\frac{n-1}{2}} & & r_{2,\frac{n+1}{2}} + S_{2,\frac{n+1}{2}} \\ & \vdots & \vdots & \vdots & \vdots \\ \vdots & \vdots & \vdots & & \vdots \\ r_{\frac{n+1}{2},1} + S_{\frac{n+1}{2},1} & \cdots & \cdots & \cdots & r_{\frac{n+1}{2},\frac{n+3}{2}} + S_{\frac{n+3}{2},\frac{n+3}{2}} \\ & \vdots & \vdots & \vdots & \vdots \\ & & r_{n-1,\frac{n-1}{2}} + S_{n-1,\frac{n-1}{2}} & & r_{n-1,\frac{n+1}{2}} + S_{n-1,\frac{n+1}{2}} \\ & & & & r_{n,\frac{n+1}{2}} + S_{n,\frac{n+1}{2}} \end{array} \right.$$

$$\left. \begin{array}{c} r_{2,\frac{n+3}{2}} + s_{2,\frac{n+3}{2}} \\[1em] \vdots \qquad \vdots \\[1em] \vdots \qquad \vdots \quad \vdots \\[1em] \cdots \qquad \cdots \quad \cdots \qquad r_{\frac{n+1}{2},n} + s_{\frac{n+1}{2},n} \\[1em] \vdots \qquad \vdots \quad \vdots \\[1em] \vdots \qquad \vdots \\[1em] r_{n-1,\frac{n+3}{2}} + s_{n-1,\frac{n+3}{2}} \end{array} \right\rangle,$$

$$\alpha R = \left\langle \begin{array}{ccccccccc} & & & & \alpha r_{1,\frac{n+1}{2}} \\ & & \alpha r_{2,\frac{n-1}{2}} & \alpha r_{2,\frac{n+1}{2}} & \alpha r_{2,\frac{n+3}{2}} \\ & & \vdots & \vdots & \vdots & \vdots & \vdots \\ & \vdots & \vdots & \vdots & \vdots & \vdots & \vdots & \vdots \\ \alpha r_{\frac{n+1}{2},1} & \alpha r_{\frac{n+1}{2},2} & \cdots & \cdots & \alpha r_{\frac{n+1}{2},\frac{n+1}{2}} & \cdots & \cdots & \alpha r_{\frac{n+1}{2},n-1} & \alpha r_{\frac{n+1}{2},n} \\ & \vdots & \vdots & \vdots & \vdots & \vdots & \vdots & \vdots \\ & & \vdots & \vdots & \vdots & \vdots & \vdots \\ & & \alpha r_{n-1,\frac{n-1}{2}} & \alpha r_{n-1,\frac{n+1}{2}} & \alpha r_{n-1,\frac{n+3}{2}} \\ & & & \alpha r_{n,\frac{n+1}{2}} \end{array} \right\rangle$$

and

A New Expression for Rhotrix

$$R \circ S = Q = \left\langle \begin{array}{ccccccccc}
 & & & & q_{1,\frac{n+1}{2}} & & & & \\
 & & & q_{2,\frac{n-1}{2}} & q_{2,\frac{n+1}{2}} & q_{2,\frac{n+3}{2}} & & & \\
 & & \vdots & \vdots & \vdots & \vdots & \vdots & & \\
 & & \vdots & \vdots & \vdots & \vdots & \vdots & & \\
q_{\frac{n+1}{2},1} & q_{\frac{n+1}{2},2} & \cdots & \cdots & q_{\frac{n+1}{2},\frac{n+1}{2}} & \cdots & \cdots & q_{\frac{n+1}{2},n-1} & q_{\frac{n+1}{2},n} \\
 & & \vdots & \vdots & \vdots & \vdots & \vdots & & \\
 & & \vdots & \vdots & \vdots & \vdots & \vdots & & \\
 & & & q_{n-1,\frac{n-1}{2}} & q_{n-1,\frac{n+1}{2}} & q_{n-1,\frac{n+3}{2}} & & & \\
 & & & & q_{n,\frac{n+1}{2}} & & & &
\end{array} \right\rangle,$$

where,

$$h(R) = r_{\frac{n+1}{2},\frac{n+1}{2}} ; \quad h(S) = s_{\frac{n+1}{2},\frac{n+1}{2}}$$

$$q_{1,\frac{n+1}{2}} = r_{1,\frac{n+1}{2}}h(S) + s_{1,\frac{n+1}{2}}h(R),$$

$$q_{2,\frac{n-1}{2}} = r_{2,\frac{n-1}{2}}h(S) + s_{2,\frac{n-1}{2}}h(R),$$

$$q_{2,\frac{n+1}{2}} = r_{2,\frac{n+1}{2}}h(S) + s_{2,\frac{n+1}{2}}h(R),$$

$$q_{2,\frac{n+3}{2}} = r_{2,\frac{n+3}{2}}h(S) + s_{2,\frac{n+3}{2}}h(R),$$

$$q_{\frac{n+1}{2},1} = r_{\frac{n+1}{2},1}h(S) + s_{\frac{n+1}{2},1}h(R),$$

$$q_{\frac{n+1}{2},2} = r_{\frac{n+1}{2},2}h(S) + s_{\frac{n+1}{2},2}h(R),$$

$$q_{\frac{n+1}{2},\frac{n+1}{2}} = r_{\frac{n+1}{2},\frac{n+1}{2}}h(S) + s_{\frac{n+1}{2},\frac{n+1}{2}}h(R),$$

$$q_{\frac{n+1}{2},n-1} = r_{\frac{n+1}{2},n-1}h(S) + s_{\frac{n+1}{2},n-1}h(R),$$

$$q_{\frac{n+1}{2},n} = r_{\frac{n+1}{2},n}h(S) + s_{\frac{n+1}{2},n}h(R),$$

$$q_{n-1,\frac{n-1}{2}} = r_{n-1,\frac{n-1}{2}}h(S) + s_{n-1,\frac{n-1}{2}}h(R),$$

$$q_{n-1,\frac{n+1}{2}} = r_{n-1,\frac{n+1}{2}}h(S) + s_{n-1,\frac{n+1}{2}}h(R),$$

$$q_{n-1,\frac{n+3}{2}} = r_{n-1,\frac{n+3}{2}}h(S) + s_{n-1,\frac{n+3}{2}}h(R),$$

and

$$q_{n,\frac{n+1}{2}} = r_{n,\frac{n+1}{2}}h(S) + s_{n,\frac{n+1}{2}}h(R).$$

The rhotrix set $\overset{\Diamond}{C}(n)$ together with above operations of addition (+), scalar multiplication (a) and multiplication (o) is a generalization of rhotrix spaces, denoted by the pair

$$\left\langle \overset{\Diamond}{C}(n), +, \circ, \alpha \right\rangle.$$

CONCLUSIONS

We have presented a statement of expression for generalization of all rhotrices of same size-n using multiple array indexes, which may be analogous to what is obtainable in matrix theory. The new rhotrix look will encourage the study of algebra and analysis of rhotrix spaces from a better perspective. In addition, computing efficiency and accuracy will also be improved, particularly when the operations over rhotrix spaces, based on the new expression are algorithmatized for computing machines. In the future research direction, it seems interesting to think of "axiomatization of real rhotrix space". This topic will be our next line of focus for research.

ACKNOWLEDGMENTS

We wish to thank Ahmadu Bello University, Zaria, Nigeria for funding this relatively new area of research.

REFERENCES

1. Ajibade, A.O. (2003) the Concept of Rhotrix for Mathematical Enrichment. International Journal of Mathematical Education in Science and Technology, 34, 175-179.http://dx.doi.org/10.1080/0020739021000053828
2. Atanassov, K.T. and Shannon, A.G. (1998) Matrix-Tertions and Matrix-Noitrets: Exercises in Mathematical Enrichment. International Journal of Mathematical Education in Science and Technology, 29, 898-903.
3. Mohammed, A. (2007) Enrichment Exercises through Extension to Rhotrices. International Journal of Mathematical Education in Science and Technology, 38, 131-136.http://dx.doi.org/10.1080/00207390600838490
4. Mohammed, A. (2007) a Note on Rhotrix Exponent Rule and Its Applications to Special Series and Polynomial Equations Defined over Rhotrices. Notes on Number Theory and Discrete Mathematics, 13, 1-15.
5. Mohammed, A. (2008) Rhotrices and Their Applicationsin Enrichment of Mathematical Algebra. Proceedings of 3rd International Conference on Mathematical Sciences, United Arab Emirate University, Alain, 1, 145-154.
6. Mohammed, A. (2009) A Remark on Classifications of Rhotrices as Abstract Structures. International Journal of Physical Sciences, 4, 496-499.
7. Mohammed, A., Ezugwu, E.A. and Sani, B. (2011) On Generalization and Algorithmatization of Heart-Based Method for Multiplication of Rhotrices. International Journal of Computer Information Systems, 2, 46-49.
8. Sani, B. (2004) an Alternative Method for Multiplication of Rhotrices. International Journal of Mathematical Education in Science and Technology, 35, 777-781.http://dx.doi.org/10.1080/00207390410001716577
9. Sani, B. (2007) the Row-Column Multiplication of High Dimensional Rhotrices. International Journal of Mathematical Education in Science and Technology, 38, 657-662.http://dx.doi.org/10.1080/00207390601035245
10. Sani, B. (2008) Conversion of a Rhotrix to a Coupled Matrix. International Journal of Mathematical Education in Science and Technology, 39, 244-249.http://dx.doi.org/10.1080/00207390701500197

CITATION

Mohammed, A. (2014) A New Expression for Rhotrix. Advances in Linear Algebra & Matrix Theory, 4, 128-133. http://dx.doi.org/10.4236/alamt.2014.42011.

Index